T0327549

STIRLING CYCLE ENGINES

STIRLING CYCLE ENGINES
INNER WORKINGS AND DESIGN

Allan J Organ

This edition first published 2014
© 2014 John Wiley & Sons, Ltd

Registered office

John Wiley & Sons Ltd, The Atrium, Southern Gate, Chichester, West Sussex, PO19 8SQ, United Kingdom

For details of our global editorial offices, for customer services and for information about how to apply for permission to reuse the copyright material in this book please see our website at www.wiley.com.

Wiley also publishes its books in a variety of electronic formats. Some content that appears in print may not be available in electronic books.

Library of Congress Cataloging-in-Publication Data

Organ, Allan J.
  Stirling cycle engines : inner workings and design / Allan J Organ.
    pages cm
  Includes bibliographical references and index.
  ISBN 978-1-118-81843-5 (cloth)
 1. Stirling engines.   2. Stirling engines–Design and construction.   I. Title.
  TJ765.O738 2014
  621.4′2–dc23

                                                    2013031868

A catalogue record for this book is available from the British Library.

ISBN: 9781118818435

Set in 10/12pt Times by Aptara Inc., New Delhi, India

1   2014

Contents

About the Author

Apart from a six-year spell developing tooling for high-speed metal forming, the author's academic research career – which concluded with 20 years in Cambridge University Engineering Department where he learned more than he taught – has focused on Stirling cycle machines.

It would be gratifying to be able to claim that this commitment had resulted in his becoming an authority – perhaps *the* authority. However, it is a feature of the Stirling engine – perhaps its irresistible attraction – that it does not yield up its subtleties quite so readily. Thus the author writes, not as a self-styled expert, but as a chronic enthusiast anxious that the results of his most recent enquiries should not expire with him! It must now fall to a more youthful intellect to pursue matters.

With the 200th anniversary of the 1816 invention fast approaching, the author remains optimistic of a commercial future for a design targeting appropriate applications and undertaken with realistic expectations. Readers interested in this vision may wish to look out for a work of fiction entitled '*The Bridge*'.

Foreword

What is the source of the Stirling cycle's siren song? Why has it attracted so many for so long (including yours truly over some 30-plus years)? In part, it is simply that Stirling engines are not in common use, so that each idealist who stumbles upon some brief description of such, perceives that he has stumbled upon hidden gold, a gift to the world that needs only to be unwrapped and presented to a grateful society that will receive its promise of clean, quiet, reliable power. No valves, no timing, no spark, no explosions – what could be simpler? Arthur C. Clarke said "Any sufficiently developed technology is indistinguishable from magic". Surely, then, with just a bit of development, or better materials, or the keen insight of one who sees what others have missed, the Stirling engine will blossom into the fruition of its magical promise.

And yet. …

The first such machines were built two centuries ago, though only retrospectively identified with their eponymous creator, Rev. Robert Stirling. That naming came mostly by virtue of his invention of the "economizer" (what we now recognize as the regenerator in a reciprocating machine of this type), and was promoted most effectively by Rolf Meijer, who led the change from air-charged machines to light-gas, pressurized devices at Philips in the 1940s. Dr. Meijer might be said to be the father of the modern Stirling engine, as the earlier work did not have the benefits of thermodynamics as a science or modern materials like stainless steel.

And yet. …

It is remarkable that some 70 years after the Philips Stirling generator sets were produced, then abandoned as unprofitable, they remain one of the benchmarks against which novel attempts at Stirling engines of practical utility are measured. Few have succeeded in bettering their technical performance, and fewer still have achieved any greater commercial returns. Countless hobbyists, dozens of corporate ventures, and even a few large-scale government projects have come and gone.

And yet. …

Not one living person in a million today has seen, used, or been empowered by a Stirling engine. Why then, do we persist in our apparently sisyphean pursuit of this esoteric system? And commensurately, why is the present book important?

It has been noted that most diversity exists at the interfaces among ecosystems: that the junctions of field and forest, sea and shore, or sky and soil support more life in more forms than the depths of any one such domain. It is sure that these are the sites of evolution. As an inventor–instructor–entrepreneur, it has been apparent to me that a similar effect applies to intellectual pursuits: the greatest opportunities for development are to be found at the intersections of disciplines, rather than at their cores. Recent advances in combination fields such as mechatronics, evolutionary biology, and astrophysics might be evidence of some truth in this observation. Let us consider the Stirling engine in this light, the better to appreciate the value of this book.

Thermodynamics, heat transfer, fluid science, metallurgy, structural mechanics, dynamics of motion: all these and more are essential elements in Stirling embodiments and their mastery serves as arrows in the quiver of the developer aiming for Stirling success. Perhaps this is one reason why we are attracted to the Stirling – each finds his own expertise essential to it. And perhaps this is one reason why success is so hard to achieve, for who has all these skills at hand? And if success demands such polymath capacities, where is one to begin?

There are many published works on the broad topic of Stirling cycles, engines, and coolers, especially if the reader seeks descriptive or historical information. A selection of analytical texts can be found for those seeking guidance on the first-principles design of aspects of a new engine, including some worthy treatises on the numerical simulation of complete engines (or coolers). Yet none provides a technically sound, computationally compact path to buildable, valid engines.

This new work builds on the author's earlier focus on the essential regenerator and the application of similarity principles, validated by well-documented machines that serve here as a basis for scaling rules and the design of new engines for applications and operating conditions that superficially differ greatly from prior examples. It must be at those new conditions, perhaps at the intersections of conventional mechanics with micro-, or bio-, or other technologies, that new and evolved implementations of Stirling technology will arise and become, perhaps, successful. In this offering, Dr. Organ does the world of Stirling developers (and would-be Stirling magnates) a great service. For many times, new energy has been brought to this field and applied without reference to the experiences, successes, and failures of the past, here applied to great effect.

That tendency to dive in without a thorough grounding in the prior art is due only in part to the aforementioned siren song of Stirling and its addling effect on the newly captured. Such repetitive waste is also driven by the relative inaccessibility of much of the greater body of Stirling technical literature (e.g., I watched the published output of one famous free-piston company go through several 2-year cycles of re-inventions, as successive tiers of graduate students rotated through!).

The challenge is that, even when accessible as correct content, much of what is published in Stirling literature is either uselessly facile or excruciatingly partial in scope, so as to preclude its ready application to new designs. Tools to fit that job have not heretofore been available to those not willing or able to amass and absorb a gargantuan (if dross-filled) library of publications and apply that through associated years of experimental training. This book, through the author's elegant nomographic presentation – fully sustained by clear text and mathematical underpinnings – provides just that holistic entry point, presented with wit and minimal pain in calculations.

Hints of whole-physics participation in Stirling analysis abound here, not least in the dismissal of Schmidt models for their gross errors and oversimplifications that have led to conceptual misunderstandings and hampered many development efforts. I am particularly pleased at the refutation of long-standing shibboleths such as the "evil" of dead space; shattered here with clear and concrete constructions of the actual effects of dead space, and its value in the right places and amounts. In my own work of recent years, which has merged a long

Stirling experience with more recently developed thermoacoustic science, the key has been full consideration of the inertial properties of the cycle fluids, which is ignored by most Stirling models and simulations. Here, those effects are illustrated and their contributions to actual Stirling device behaviors discussed in a unique bridge between closed-form, analytical methods, and the full physics of numerical simulation, including a proper dressing-down and reformulation of the steady-flow correlations so often misapplied to this oscillatory system. The resulting graphics are both useful and beautiful.

Dr. Organ's offered tools, *FastTrack, FlexiScale, and ReScale* fulfill his promise of guiding the designer of a new Stirling engine to a safe island in the sea of possibilities. This traceable relation from technically successful engines of the past (although I am, of course, crushed that none of mine are among those cited), without the need to extract and refine the data from disparate sources elsewhere, opens the possibility of building a useful Stirling engine to a much larger population of aspirants. Perhaps by this means, some clever member thereof will at last find the sweet spot for commercial success; but at the very least, innumerable hours that would otherwise have been wasted in blind stabs can now be channelled into production refinements on a sturdy base. This is indeed a grand achievement, and being provided in so readable a volume is all the more so: a gift to the Stirling Community sure to be acclaimed throughout.

I am honored to call this author my friend and fellow explorer, and to introduce you, the reader, to this work with the certainty that if you have heard already that siren song of Stirling, this book by Allan Organ can lead you to safe harbor in plotting your course in response. Gentlefolk, Start Your Engines!

Dr John Corey

Preface

If the academic study of the Stirling engine began with Gustav Schmidt in 1861, then it has been more than a century and a half in the making. This might be deemed more than sufficient time to achieve its purpose – which must surely be to put itself out of a job.

A symbolic date looms: Tuesday 27 September 2016 – the bi-centenary of Stirling's application for his first patent – for his 'economizer' (regenerator). The Stirling engine will by then have been under development – admittedly intermittent development – for two centuries. Yet a would-be designer continues to be faced with the unhappy choice between (a) proceeding by trial and error and (b) design from thermodynamic first principles. Either course can be of indeterminate cost and duration.

In principle, nothing stands in the way of an approach to thermodynamic design which is both general and at the same time 'frozen' – general in the sense of coping with arbitrary operating conditions (*rpm*, charge pressure, working gas) and 'frozen' to the extent that thermodynamic design (numbers, lengths and cross-sectional dimensions of flow channels, etc.) are read from graphs or charts, or acquired by keying operating conditions and required performance into a lap-top computer or mobile phone 'app'. The possibility arises because, from engine to engine, the gas process interactions by which heat is converted to work have a high degree of *intrinsic similarity*. Physical processes which are formally similar are *scalable*: once adequately understood and rendered in terms of the appropriate parameters, no compelling reason remains for ever visiting them again.

Market prospects must surely be improved by relegating the most inscrutable – and arguably most daunting – aspect of the design of a new engine to a few minutes' work.

The present account is motivated by a vision along the foregoing lines. Utopia remains on the horizon, but there is progress to report:

> Wherever possible, working equations are reduced to three- and four-scale nomograms. The format affords better resolution and higher precision than the traditional $x - y$ plot, and allows a range of design options to be scanned visually in less time than it takes to launch equivalent software on a computer.
>
> New, independently formulated algorithms for thermodynamic scaling endorse the original method (*Scalit*) and increase confidence in this empirical approach to gas path design. The scaling sequence now reduces to the use of nomograms.
>
> There is a novel – and unprecedentedly simple – way of inferring loss per cycle incurred in converting net heat input to indicated work.
>
> Steady-flow heat transfer and friction correlations appear increasingly irrelevant to conditions in the Stirling engine. Attention shifts to the possibility of correlating *specific thermal load* per

tube against a *Reynolds parameter* for the multi-tube exchanger assembly tested *in situ*. Results are promising. Given that they derive from Stirling engines under test, relevance to the unsteady flow conditions is beyond question.

Progress has resulted from noting that the context does not call for a comprehensive picture of regenerator transient response: the interests of the mathematician do not coincide with the realities and requirements of satisfactory engine operation. Focusing on the relevant margin of the potential operating envelope allows respective temperature excursions of gas and matrix to be explored independently.

The kinetic theory of gases is mobilized in an attempt to dispense once for all with the suspect steady-flow flow correlations. To convey the resulting insights calls for animated display. This is not yet a feature of the conventional hard-copy volume. Selected still frames give an impression.

(Full exploitation of the kinetic theory formulation awaits the next generation of computers, so another book looms. Another book – another preface!)

Certain entrenched tenets of thermodynamic design have been found to be faulty. These are remedied.

Versatility and utility of the design charts (nomograms) have been enhanced: The range of equations susceptible to traditional methods of construction is limited. The technique of 'anamorphic transformation' has been corrected relative to published accounts, and now allows display of functions of the form $w = f(u,v)$ in nomogram form. Function w can be the result of lengthy numerical computation. (The display remains confined to the range over which the variation in w is monotonic.)

The 'hot-air' engine receives a measure of long-overdue attention.

The writer has benefited from long hours of dialogue with Peter Feulner, with Geoff Vaizey, with Camille van Rijn, with Peter Maeckel and with R G 'Jimmy' James. The influence of Ted Finkelstein endures.

Constructive criticism is always welcome at *allan.j.o<at>btinternet.com*.

This material appears in print thanks largely to a unique combination of persistence, patience and diplomacy applied over a period of 18 months by Eric Willner, executive commisioning editor at John Wiley & Sons. The project has since become reality in the hands of Anne Hunt, associate commissioning editor and Tom Carter, project editor.

Notation

Variables Having Dimensions

A_{ff}	free-flow area	m²
A_w	wetted area	m²
b	width of slot	m
c_v, c_p	specific heat at constant volume, pressure	J/kgK
d_x	internal diameter of heat exchanger duct	m
D	inside diameter of cylinder, or outside diameter of displacer, as per context	m
e	désaxé offset	m
f	cycle frequency ($= \omega/2\pi$)	s⁻¹
g	radial width of annular gap	m
h	coefficient of convective heat transfer	W/m² K
	specific enthalpy	J/kgK
H	enthalpy	J
H_C	clearance height	m
k	thermal conductivity	W/mK
L_d	axial length displacer shell	m
L_{ref}	reference length $V_{sw}^{1/3}$	m
L_x	length of heat exchanger duct	m
m	mass (of gas)	kg
M	total mass of gas taking part in cycle	kg
M_w	mass of matrix material	kg
p_{ref}	reference pressure – max/min/mean cycle value	Pa
p_w	wetted perimeter	m
Q	heat	J
r	linear distance coordinate in radial direction	m
	radius – e.g., of crank-pin offset	m
r_h	hydraulic radius	m
q'	heat rate	W
q''	heat rate per unit length of exchanger	W/m
R	specific gas constant	J/kgK
S_p, S_d	stroke of work piston and displacer respectively	m
T_E, T_C	temperatures of heat source and sink	K

T_w	temperature of solid surface	K
T	temperature of gas	K
t	time	s
	thickness in radial coordinate direction	m
u	velocity in x coordinate direction	m/s
	specific internal energy	J/kgK
U	internal energy	J
V_{sw}	swept volume	m^3
W	work	J
W'	work rate	W
X, Y	linear distances in x and y coordinate directions	m
z	linear offset in kinematics of crank-slider mechanism	m
α	thermal diffusivity $k/\rho c_p$	m^2/s
ε_T	mean temperature perturbation or 'error' $\overline{T - T_w}$	K
μ	coefficient of dynamic viscosity	Pas
ρ	density	kg/m^3
ω	angular speed	s^{-1}

Dimensionless Variables

a, b, c, d	coefficients and indices as required	-
a	coefficients of linear algebraic equations	-
C	numerical constant (as required)	-
CI	'cycle invariant' defined at point of use	-
C_f	friction factor $\Delta p/\frac{1}{2}\rho u^2$	-
DG	dimensionless group defined at point of use	-
Ma	Mach number $u/\sqrt{\gamma RT}$	-
n	polytropic index: n_e expansion phase; n_c expansion phase	-
n_{Tx}	number of exchanger tubes	-
N_B	specific cycle work: power/$fV_{sw}p_{ref}$	-
N_F	Fourier modulus $\alpha/\omega r_0^2$	-
N_{FL}	Flush ratio: ratio of mass of gas per uni-directional blow to instantaneous mass of gas in regenerator void volume.	-
N_{MA}	characteristic Mach number $\omega L_{ref}/\sqrt{RT_C}$	-
N_{Nu}	Nusselt number hr_h/k	-
N_{RE}	characteristic Reynolds number $N_{SG}N_{MA}^2$	-
N_{SG}	Stirling parameter $p_{ref}/\mu\omega$	-
N_T	characteristic temperature ratio T_E/T_C	-
N_{TCR}	thermal capacity ratio $\rho_w c_w T_C/p_{ref}$	-
NTU	$\underline{N}$umber of $\underline{T}$ransfer $\underline{U}$nits $StL_x/r_h = hT_C/\omega p_{ref}S$ in Carnot cycle study	-
P	parameter in eq'n. which relates L_x/d_x to L_x/L_{ref}	-
$P(\varphi), Q(\varphi)$	consolidated coefficients of first-order differential equation	-
Pr	Prandtl number $\mu c_p/k$	-
QI	'quasi-invariant' defined at point of use	-
r_v	volumetric compression ratio (e.g., V_1/V_3 of Carnot cycle)	-
r_p	pressure ratio p_{max}/p_{min}	-

Re	Reynolds number $4\rho\lvert u\rvert\mu/r_\mathrm{h}$	-
RE_ω	Reynolds number characteristic of exchanger operation over a cycle	-
S	linear scale factor $L_\mathrm{d}/L_\mathrm{p} = L_\mathrm{derivative}/L_\mathrm{prototype}$	-
Sg	Stirling number $pr_\mathrm{h}/\mu\lvert u\rvert$ (see Stirling *parameter* above)	-
St	Stanton number $h/\rho\lvert u\rvert c_\mathrm{p}$	-
TCR	net thermal capacity ratio (Chapter 17)	-
$\mathrm{U(\ )}$	step function used in ideal adiabatic cycle	-
x	numerical scale factor	-
x, y, z	cartesian coordinates	-
XQ_x	specific thermal load on exchanger assembly	-
XQ_XT	specific thermal load on individual tube of exchanger assembly	-
Z	work quantity (e.g., loss per cycle) normalized by MRT_C or by $p_\mathrm{ref}V_\mathrm{sw}$	-
α	phase advance of events in expansion space over those in compression space	-
β	phase advance of displacer motion over that of work piston	-
γ	isentropic index – specific heat ratio	-
Δ	any finite difference or change	-
ΔT	temperature difference $T - T_\mathrm{w}$	-
δ	dimensionless dead space $V_\mathrm{d}/V_\mathrm{sw}$	-
φ	crank angle $= \omega t$	-
θ	angular coordinate in circumferential direction	-
	angle through which coordinate frame is rotated in process of dealing with molecular collision	-
κ	thermodynamic volume ratio $V_\mathrm{C}/V_\mathrm{E}$	-
λ	ratio of volume swept by work piston to that swept displacer $\approx S_\mathrm{p}/S_\mathrm{d}$ in parallel-cylinder, coaxial 'beta' machine	-
Λ	Hausen's 'reduced length' – equivalent to NTU	-
ν	Finkelstein's dimensionless dead space $2\Sigma\upsilon_i T_i/T_\mathrm{C}$	-
ψ	specific pressure p/p_ref	-
Π	Hausen's 'reduced period' $hA_\mathrm{w}L_\mathrm{r}/fM_\mathrm{w}c_\mathrm{w}$	-
ρ	composite dimensionless function arising in Finkelstein's formulation of Schmidt analysis: $\sqrt{\{N_\mathrm{T}^{-2} + \kappa^2 + 2\kappa\cos(\alpha)/N_\mathrm{T}\}}/(N_\mathrm{T}^{-1} + \kappa + \nu)$	-
σ	specific mass $mRT_\mathrm{C}/p_\mathrm{ref}V_\mathrm{sw}$	-
σ'	specific mass rate $d\sigma/d\varphi = mRT_\mathrm{C}/\omega p_\mathrm{ref}V_\mathrm{sw}$	-
ς	dimensionless length x/L_ref	-
τ	Finkelstein's temperature ratio $T_\mathrm{C}/T_\mathrm{E} = N_\mathrm{T}^{-1}$	-
υ	Finkelstein's proportional dead space $V_\mathrm{di}/V_\mathrm{E}$	-
Σ	total inventory of specific mass $MRT_\mathrm{C}/p_\mathrm{ref}V_\mathrm{sw} = \sigma_\mathrm{e} + \sigma_\mathrm{c} + \nu\psi$ – assumed invariant	-
$\P_\mathrm{v}$	volume porosity	-

Subscripts

comb	combustion
C	relating to compression or to compression space
d	relating to displacer

env   'envelope' of overlapping displacements of piston and displacer in coaxial,
           'beta' machine
exp   experimental
E   relating to expansion or to expansion space
f   relating to flow friction
ff   free-flow (area)
i   in or inlet
j,k   array subscripts
L   lost – as in lost available work
o   out or outlet
p   relating to piston
q   relating to heat transfer
r   regenerator
ref   reference value of variable
rej   (heat) rejected
sw   swept (volume)
w   wetted (area), wall or wire, as per context
x   exchanger: $_{xe}$ – expansion exchanger; $_{xc}$ – compression exchanger
∞   free-stream

Superscripts

deriv   derivative design of scaling process
prot   relating to specification of prototype of scaling process
T   relating to the new (transposed) reference frame
+   extra or additional (dead space)

1

Stirling myth – and Stirling reality

1.1 Expectation

'*Stirling's is a perfect engine, and is the first perfect engine ever to be described.*' (Fleeming Jenkin, 1884). '*offering silence, long life.*' (Ross, 1977). '*... thus enabling the thermal efficiency of the cycle to approach the limiting Carnot efficiency.*' (Wikipedia 2013). *A silent, burn-anything, mechanically simple, low-maintenance, low-pollution prime mover with potential for the thermal efficiency of the Carnot cycle.* Here, without doubt, is a recipe for run-away commercial success which Lloyds of London would surely be happy to underwrite.

The reality of the modern Stirling engine is in terms of tens of thousands of 'one-offs' – prototypes or designs of different degrees of sophistication, only a tiny handful of which have been followed up by a degree of further development.

The outcome of a technological venture has much to do with expectation: Where this is unreasonably or irrationally high, the outcome falls short. The verdict – by definition – is failure. The *identical* technological outcome based on realistic expectation can amount to success.

The Stirling engine is not silent – but can be quiet relative to reciprocating internal combustion engines of comparable shaft power. There is not the remotest chance of approaching the so-called Carnot efficiency, but claims for brake thermal efficiencies comparable to those achieved by the diesel engine appear genuine. Nominal parts count per cylinder of the multi-cylinder Stirling is, indeed, lower than that of the corresponding four-stroke IC engine. On the other hand, many individual components pose a severe challenge to mass manufacture.

Supposing this more sober view to be correct, a world of limited resources and increasing environmental awareness probably has room for the Stirling engine. If so, responding to the need is going to require a lowering of expectation. This may be helped by the shedding of a substantial body of myth.

Stirling Cycle Engines: Inner Workings and Design, First Edition. Allan J Organ.
© 2014 John Wiley & Sons, Ltd. Published 2014 by John Wiley & Sons, Ltd.

1.2 Myth by myth

1.2.1 That the quarry engine of 1818 developed 2 hp

This can be traced back to an article in *The Engineer* of 1917 celebrating rediscovery of the patent specification – but no further. A back of envelope calculation will indicate whether further enquiry is necessary.

All power-producing Stirling engines of documented performance yield approximately the same value of Beale number N_B:

$$N_B = \frac{\text{shaft power [W]}}{\text{charge pressure [Pa]} \times \text{swept volume [m}^3\text{] and speed [cps} = rpm/60\text{]}}$$

The value is dimensionless and is typically 0.15 [–] – the 'Beale number'.

Charge pressure p_{ref} of Stirling's engine was 1 atm or 10^5 Pa. Linear dimensions cited in the 1816 patent convert to swept volume V_{sw} of 0.103 m^3. Rotational speed is not on record, but on the basis of hoop stress, the *rpm* capability of the 10-ft diameter composite flywheel has been estimated with some confidence (and with subsequent corroboration – Organ 2007) at 27.

On this basis the Beale arithmetic suggests a power output of 695 W – or 0.93 hp. The figure of 0.15, however, derives from engines operating with expansion-space temperatures T_E at or above 900 K (627 °C). Stirling is specific about a much lower value of 480 °F, consistent with limitations of materials available in 1818 (wrought iron). This converts to 297 °C, or 570 K. Performance is very much a function of temperature ratio, $N_T = T_E/T_C$, where T_E and T_C are absolute temperatures, K. Where N_T departs from the norm of $N_T = N_T{}^* = 3$, the definitive parametric cycle analysis (Finkelstein 1960a) justifies a temperature correction factor of $(1 - 1/N_T)/(1 - 1/N_T{}^*)$. The factor is 0.345/0.666 in the present instance, or 0.518, reducing the shaft power prediction to 0.482 hp, or 360 W.

The figure is corroborated by a forensic study (Organ 2007) based on experiments with a full-size replica (Figure 1.1) of furnace and displacer cylinder, backed up by computer simulation.

Chapter 1 of the 2007 text describes the construction of the replica furnace and flue, the stoking experiments and temperature measurements. At maximum stoking rate on coal, peak temperature measured at the rectangular exhaust outlet is 200 °C. It is possible to hold a hand against the upper inner surface of the expansion cylinder, suggesting that the internal temperature of the metal surface remains below 60 °C.

If the inevitable conclusion fails to convince, then it may help to recall the last time an egg was successfully boiled with the heat source some 10 feet (3 m) from the saucepan.

The simulation re-created the cyclic volume variations generated by the crank mechanism of the 1816 patent drawings. The start-point for exploratory simulation runs was the earlier estimate of flywheel-limiting speed of 27 *rpm* derived from considerations of hoop stress. Interpretation of the results erred in favour of best possible performance: Wire diameter d_w and winding pitch (essentially inverse of mesh number m_w) settled on for the regenerator were those which caused the simulation to indicate optimum balance between loss due to pumping and loss due to heat transfer deficit.

On this basis peak power occurred at 28 *rpm*. The simulation could coax no more than ½ hp (373 W) from the crankshaft.

Figure 1.1 Furnace, flue and upper displacer cylinder re-constructed full-size to the drawings and dimensions of the Edinburgh patent

Stirling's own hand-written description cites temperature *difference* (between upper and lower extremes of the cylinder) of 480 °F – or 297 °C. If ambient temperature were 30 °C this would require the hottest part of the cylinder to be at 337 °C. The quarry engine doubtless functioned – but not in the elegant configuration of the patent drawing.

A drawing of an engine which, by contrast, is readily reconciled with brother James Stirling's retrospective (1852) account is shown at Figure 1.2: flywheel, link mechanism, cylinder, piston, and displacer are re-used. However, the entire assembly now operates upside down relative to the patent illustrations, with cylinder head immediately above the furnace and the flame in direct contact.

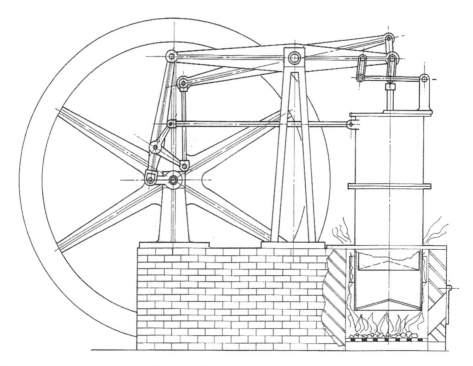

Figure 1.2 How the quarry engine might have appeared after inversion of the cylinder unit to allow heating to a viable operating temperature

Achieving ½ hp no longer requires Stirling to have optimized the thermal and flow design of the regenerator. Brother James' account of the eventual failure of the engine now makes sense: '…*the **bottom** of the air vessel became over-heated.*'

1.2.2 That the limiting efficiency of the stirling engine is that of the Carnot cycle

Nothing could be further from the truth! The Stirling engine functions by virtue of an irre-versible process – that of of forced, convective heat transfer. It does so in spite of two further irreversibilities: thermal diffusion and viscous fluid flow. The *engine* – the hardware as distinct from the academic distraction – no more aspires to the Carnot efficiency than does the cement mixer.

The spurious comparison raises important matters which will need re-visiting at a later stage:

> The Carnot cycle appears to be imperfectly understood – not least by the man himself, Sadi Carnot.
> A part-understood criterion is not one which gets applied logically.

(There is something mildly fraudulent afoot when a principle hailed as an unsurpassed ideal on the blackboard promises dismal failure when converted to hardware. A Carnot 'engine' would be crippled by inadequate thermal diffusivity of the working gas, or by sealing problems – or by both. When *any* Stirling engine turns a crank, thermal efficiency and specific power surpass those of the hypothetical Carnot embodiment.)

Used appropriately, however, the ideal cycle has unexpected insight to offer: the element of net work W [J] when heat Q_E [J] is admitted to the cycle at T_E [K] is $Q_E(T_E - T_C)/T_E$ [J]. The value reflects the absence of losses of any kind. When the practical Stirling engine takes in Q_E per cycle at T_E, the resulting W is less than the Carnot value to the extent of net losses. For a given engine the numerical total of the loss per cycle varies with charge pressure, *rpm*, and so on. The point is, however, that the value of that net loss is given by simple numerical subtraction, and that a figure is available corresponding to each documented combination of charge pressure, *rpm* and working fluid.

Chapter 3 exploits this approach. The sheer magnitude of the loss relative to the useful work element will probably come as a revelation.

1.2.3   That the 1818 engine operated '… on a principle entirely new'

There is no doubting Stirling's integrity in making the claim, but it conflicts with a reality of some 22 years earlier. The patent granted in 1794 to Thomas Mead describes '*Certain methods … sufficient to put and continue in motion any kind of machinery to which they may be applied.*' Mead describes a pair of mating cylinders coaxial on a common vertical axis. The cylinder diameters overlap (cf. the telescope), so axial motion of one relative to the other changes the enclosed volume. In one of his several 'embodiments', the outer ends of both are closed, except for provision for a rod to pass through a gland in the lowermost closure and to actuate a cylindrical 'transferrer'. Heating the top of the uppermost cylinder from the outside and alternately raising and lowering the 'transferrer' promises a cyclic swing in pressure. Mead anticipated that this would cause the 'telescope' alternately to shorten and lengthen, affording reciprocating motion capable of being harnessed.

The Mead specification, eventually published in printed form in 1856, is illustrated by a line drawing ('scetch') devoid of crank mechanism and making no pretence at depicting engineering reality. Whether or not Robert Stirling was aware of the patent, here nevertheless, in 1794, had been the essence of the closed-cycle heat engine with displacer.

1.2.4   That the invention was catalyzed by Stirling's concern over steam boiler explosions

Authors promoting this version of Stirling's motivation have been contacted. Those who have replied have declared themselves unable to offer evidence from written record.

Flynn et al. note that statistics on steam boiler explosions for the early nineteenth century are lacking. Figures attributed to Hartford Boiler Insurance Co. suggest a heyday of explosions in England – 10 000-plus incidents – during the period between 1862 and 1879. This is half a century *after* the 1816 invention.

The comprehensively-researched account by 1995 by Robert Sier of the life and times of Robert Stirling gains much of its scholarly impact by offering a wealth of quotation from the historical record – and by withholding speculation about motives.

The United Kingdom considers Frank Whittle to be the inventor of the jet engine. As a Royal Air Force officer, Whittle would have been aware of the history of gruesome amputations by aircraft propellors. No one has (so far) had the bad taste to offer that awareness in explanation of his pioneering work.

1.2.5 That younger brother James was the true inventor

The younger sibling would have been 16 at the date of filing – and only 15 while Robert was incubating the invention. Robert was living in Kilmarnock, while the Stirling family home (home of the un-married James) was 115 km distant as the crow flies (70 miles) in Perthshire.

This does not deter Kolin from asserting: '... *invention of the Stirling engine is generally attributed to Robert Stirling, which may be attractive, but is rather doubtful.'* ... *'The only written sign that Robert was engaged on the Stirling engine is his name, together with his brother James* (!) *on the patent specifications.'* (Kolin, 2000).

The statement is at odds with the meticulous attention to detail which distinguishes other work by Kolin. The patent description is in the name of Robert (alone), in his handwriting, and over his signature and seal. Evidence for a precocious contribution from James is, at the time of this writing, non-existent.

1.2.6 That 90 degrees and unity respectively are acceptable 'default' values for thermodynamic phase angle α and volume ratio κ

The false belief accompanies – and may stem from – a view that performance is more sensitive to changes in gas path specification (hydraulic radius, flow-passage length) than to changes in phase angle and volume ratio.

Systematic and long-overdue bench tests on a 'Vari-Engine' designed and built by Larque with input from Vaizey now convincingly indicate otherwise. If volume variation is simple-harmonic, $\alpha = 90$ degrees converts to $\beta = 45$ degrees and $\kappa = 1.0$ to $\lambda = 1.4$ of the 'beta' or coaxial configuration. (This is pure trigonometry – see Chapter 4 *Kinematics*. Thermodynamic considerations do not arise!) The ratio 1.4 is almost precisely the *inverse* of the value since demonstrated experimentally by Larque to give best operation of a small but 'real' engine of beta configuration (multi-channel exchangers, foil regenerator, modest pressurization on air).

Converting Larque's optimum values to the kinematics of the 'alpha' (opposed-piston) configuration yields $\alpha = 132$ degrees. and $\kappa = 0.78$. Both figures are readily embodied into a one cylinder-pair unit, but only the latter can be achieved in the four-cylinder, double-acting 'Rinia' configuration. In any case a better choice for the Rinia would now appear to be three cylinders with $\alpha = 120$ degrees – or six cylinders with an inter-connection between alternate pairs.

1.2.7 That dead space (un-swept volume) is a necessary evil

At *rpm* and pressure giving viable operation, the gas processes in the variable-volume spaces are closer to adiabatic than to isothermal. Chapter 4 confirms, amplifies and argues that, one this basis the cycle of the regenerative gas turbine (Joule cycle – adiabatic, constant pressure) throws more useful light on the operation of the practical Stirling engine than does the Schmidt cycle with its assumption of isothermal processes.

Thermal efficiency of the regenerative gas turbine reaches a peak as pressure ratio increases to relatively modest values, and falls off thereafter. Dead space in the Stirling engine is linked to pressure ratio. Pressure ratio is implicated in mechanical friction and in seal blow-by (a phenomenon not to be under-estimated) and, crucially, in the adiabatic component of temperature swing. At any given combination of temperature ratio N_T, volume phase angle α and volume ratio κ, pressure ratio increases as dead space decreases, and vice versa. Finkelstein's generalization '... *harmful dead space to be minimized* ...' may be the one and only over-simplification for which he can be criticized.

A related fallacy is that reduced charge pressure can be offset by increased pressure ratio. The adiabatic heating component is unaffected by the former, but strongly influenced by the latter. If pressure ratio were to increase in response to more effective separation of compression and expansion phases, then the net result might be performance improvement – but the matter appears not to have been explored.

1.3 ... and some heresy

Items of myth addressed to this point are anodyne romantic notions compared to the real problem: perverse application of heat transfer and flow correlations acquired from steady, incompressible flow to flow which is essentially unsteady and compressible. The problem pervades computer simulation as well as some aspects of scaling.

The matter will warrant a dedicated chapter. Meanwhile it is sufficient to note that steady-flow correlations most widely applied to regenerator calculations were not acquired from tests on regenerators. Moreover, high Reynolds numbers were achieved by sleight of hand: not at high flow speeds through fine mesh, but by low speeds through crossed rods of massive diameter. The handy short-cut masked compressibility effects now suspected of limiting the *rpm* of engines charged with heavy gases – air or nitrogen.

Dealing satisfactorily with this chronic problem would require systematic experimental work. If it is undertaken at all it will doubtless take decades. The delay must not be allowed to get in the way of progress on other fronts. Where the algebra of these chapters would otherwise be unable to proceed for lack of appropriate data, formulation will be such that the offending correlations are incorporated as 'place-holders' pending eventual replacement by legitimate plug-ins.

1.4 Why this crusade?

Left uncontested, dis-information eventually subverts reality. Regrettably, the process has already built up a head of steam: The banner of web-site www.discoverthis.com is '*Educational Science*'. At 26 February 2012 the contribution by educationalist Kathy Bazan reads: '*that's exactly what the Stirling engine is; it's an engine created for compassionate reasons.*' In Ms. Bazan's fantasy world of 1816 '*steam engines powered everything from the smallest household device* (steam-powered blender?) *to the largest steam ships and trains. Yes, people were able to get from London to Cardiff quicker than ever before ...*' (Cardiff station opened in 1850, Stephenson's *Rocket* not having run until 1829.)

Stirling's inventions (regenerator and engine) trump those of more celebrated names: While Watt, Whittle, and others built on pre-existing technology emanant in tangible, functioning hardware, the study of thermodynamics was not around to guide Stirling. As far as is known, no Mead 'engine' ever turned a crank. A precedent for the regenerator principle has yet to be suggested.

The un-sung genius of Robert Stirling merits the attention of scholarship rather than of creative journalism.

2

Réflexions sur le cicle de Carnot

2.1 Background

There are few written accounts of the Stirling engine which fail to mention the ideal Carnot[1] cycle. The purpose of inclusion is to compare its efficiency $\eta_C = 1 - T_C/T_E$ with the efficiency potential of the practical Stirling engine. Chapter 1 has already drawn attention to a spurious comparison. If the objection needed strengthening:

> The Stirling engine is a viable prime mover. The Carnot cycle is an abstraction which has yet to demonstrate that it can make a living turning a crank.
>
> It is difficult to conceive of a practical embodiment of the Carnot cycle having any chance of approaching *its own* limiting efficiency. Constructed from conventional materials it would be crippled by thermal diffusion or by sealing problems – or by both.

The original proposition (Carnot 1824) envisages a reciprocating embodiment. The gas process sequence described by Carnot does not define a closed cycle. Follow-up accounts overlook this aspect, dwelling instead on the indicator diagram. The latter has gained a reputation for an unfavourable ratio of p-V area to peak pressure. There is, however, no unique indicator diagram, which instead is a function of no fewer than three independent dimensionless parameters: temperature ratio $N_T = T_E/T_C$, specific heat ratio γ, and compression ratio $r_v = V_1/V_3$. Shape and mean effective pressure vary markedly depending on the combination of numerical values.

This writer has yet to come across a second-hand account of the Carnot cycle referring to anything deeper than the usual arbitrary quartet of intersecting isotherms and adiabats. Putting the Carnot/Stirling comparison onto a sounder footing will mean re-visiting the ideal Carnot cycle. Taking things a step closer to the reality of a Carnot 'engine' will have insights for other externally heated reciprocating prime movers – Stirling, thermal lag, and so on.

---

[1]For present purposes this means the *reciprocating* embodiment of Carnot's original account.

Stirling Cycle Engines: Inner Workings and Design, First Edition. Allan J Organ.
© 2014 John Wiley & Sons, Ltd. Published 2014 by John Wiley & Sons, Ltd.

2.2 Carnot re-visited

The ideal Carnot cycle is universally taken for granted. It is re-stated here (in Appendix A-1) firstly to rectify the deficiency of the original *Réflexions*, and secondly to introduce the reality that the volume ranges occupied by the respective phases – isothermal and adiabatic – are not arbitrary, but require to be pre-set in terms of temperature ratio N_T, compression ratio r_v, and isentropic index γ. With reference to Figure A-1.1 (Appendix A-1):

Initial compression from V_1 to V_2 is isothermal:

$$T_2 = T_1 \tag{2.1}$$

The adiabatic phase between volumes V_2 and V_3 can be defined:

$$T_3/T_2 = (V_2/V_3)^{\gamma-1} \tag{2.2}$$

Combining Equations 2.1 and 2.2:

$$(T_3/T_1)^{1/(\gamma-1)} = (V_2/V_1)(V_1/V_3)$$

The term V_1/V_3 is volumetric compression ratio r_v, and T_3/T_1 is characteristic temperature ratio N_T:

$$V_2/V_1 = r_v^{-1}N_T^{1/(\gamma-1)} \tag{2.3}$$

A viable cycle requires $V_2 < V_1$, a condition which an arbitrary combination of N_T, r_v, and γ does not necessarily satisfy. The selected indicator diagrams illustrate: all p-V loops of Figure 2.1 are for a common temperature ratio N_T, but successive loops are for increasing compression ratio r_v. The higher the value of r_v the greater is the isothermal volume change (i.e., the smaller is volume ratio V_2/V_1) required to prevent the adiabatic phase overshooting target temperature T_3.

By similar reasoning for the adiabatic expansion between volumes V_4 and V_1:

$$V_1/V_4 = r_v N_T^{-1/(\gamma-1)} \tag{2.4}$$

Values of V_2/V_1 required when N_T varies at given compression ratio r_v are less self-evident, but can be picked off Figure 2.2, which is the nomogram equivalent of Equation 2.3. A straight line cutting the central scale at an N_T value greater than unity joins viable combinations of r_v and V_2/V_1.

An unpromising feature of the cycle is the low mean effective pressure in relation to peak pressure (slenderness of the indicator diagram). Figure 2.1 nevertheless shows this feature improving somewhat with increasing compression ratio. The extent of the improvement is seen more clearly in Figure 2.3 – the temperature–entropy counterpart of Figure 2.1.

The closed sequence of process paths may be thought of as a 'static' cycle. Isothermal phases based on a real gas require infinite time, realizable only at zero speed. A cycle of real processes is inherently speed-dependent, a feature probed here by replacing both isothermal phases with processes in which instantaneous rate of heat exchange is determined by a fixed

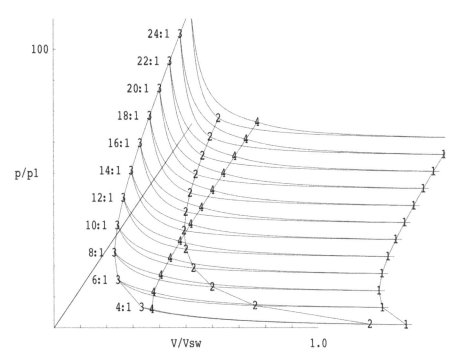

Figure 2.1   Specimen indicator diagrams for temperature ratio $N_T = T_3/T_1 = 500/300$. Parameter is compression ratio $r_v = V_1/V_3$

value of h – coefficient of convective heat transfer [W/m^{2}K]. There are unexpected findings, which will be easier to explain for being introduced through a special case.

2.3   Isothermal cylinder

A friction-less piston seals a fixed mass M of ideal gas within a closed cylinder of internal diameter D. In anticipation of extension to the Carnot configuration, piston head and cylindrical walls are adiabatic. Only the cylinder head at temperature T_C exchanges heat with the gas, doing so at an instantaneous rate determined by varying temperature difference $\Delta T = T - T_C$ and coefficient of heat transfer h. The latter can can take any value between zero (adiabatic case) and infinity (isothermal), but the numerical value is constant while the variation of T is computed over a succession of complete cycles.

Variation of volume V can be any desired function of time t (and crank angle $\varphi = \omega t$). The scope is illustrated by assuming the variation to be simple-harmonic. The piston covers stroke S while sweeping volume V_{sw} between a maximum of V_1 and minimum of V_3.

The energy equation corresponding to elemental volume change dV is:

$$\mathrm{d}Q - p\mathrm{d}V = \mathrm{d}U$$

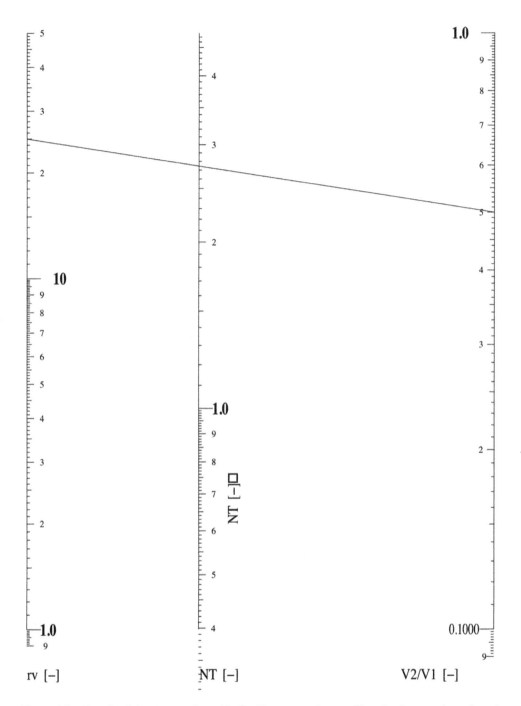

Figure 2.2 Equation 2.3 represented graphically. Nomogram format offers simultaneous inversion of the original equation

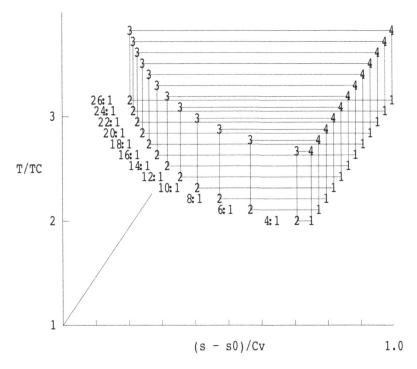

Figure 2.3 *T-s* counterpart of *p-V* diagrams of Figure 2.1. Parameter: compression ratio r_v

At any instant, temperature T is assumed uniform throughout mass M of gas, allowing internal energy U to be expressed as Mc_vT. With A_w for wetted area (area of cylinder head):

$$hA_w(T_C - T) - p\,dV/dt = Mc_v\,dT/dt \tag{2.5}$$

Pressure p is replaced by MRT/V (ideal gas law) and R/c_v by $\gamma - 1$. T can be manipulated as $T - T_C + T_C$. Crank angle $\varphi = \omega t$, so that ωdt has become $d\varphi$. Finally, $(T - T_C)/T_C$ is abbreviated to dimensionless temperature difference $\Delta\tau$. Re-arranging then gives:

$$(\gamma - 1)\frac{hA_wT_C}{\omega p_{ref}V_{sw}}\Delta\tau + (\gamma - 1)\Delta\tau V^{-1}dV/d\varphi + d\Delta\tau/d\varphi = (\gamma - 1)V^{-1}dV/d\varphi$$

The term containing h has the character of *NTU* from steady-flow heat transfer. Under the present assumptions the ratio $V_{sw}/A_w = (\frac{1}{4}\pi D^2 S)/(\frac{1}{4}\pi D^2) = S$

$$NTU = \frac{hT_C}{\omega p_{ref}S} \tag{2.6}$$

Re-arranging:

$$d\Delta\tau/d\varphi + \Delta\tau(\gamma - 1)\{NTU + V^{-1}dV/d\varphi\} = -(\gamma - 1)\tau_w V^{-1}dV/d\varphi \tag{2.7}$$

The equation is evidently of the form $d\Delta\tau/d\varphi + \Delta\tau P(\varphi) = Q(\varphi)$:

$$P = NTU + V^{-1}dV/d\varphi$$
$$Q = -(\gamma - 1)\tau_w V^{-1}dV/d\varphi$$

It is already possible to anticipate features of the eventual temperature solutions:

In the definition of P, the term $V^{-1}dV/d\varphi$ changes sign between compression and expansion. NTU is a modulus, and always positive. Despite symmetry of the simple-harmonic volume variations, a plot of T versus φ will be asymmetric.

The history of of p with φ will similarly be asymmetric: a plot of p versus V for the compression stroke will not generally superimpose over that for expansion. The result will be a loop which, for P other than zero or infinite will enclose a finite (negative) area.

Substituting the symmetrical volume variations by, for example, a crank/connecting-rod system having désaxé offset e (as in Appendix A-1), accentuates the asymmetry of the gas cycle with one combination of rotation and offset, and works against it when either e or rotation is reversed in sign.

Over a limited range of NTU, parameter P passes through zero twice per cycle. For an instant Equation 2.7 reduces to

$$\Delta\tau_{\varphi+\Delta_\varphi} - \Delta\tau_\varphi \approx \tau_w(\gamma - 1)dV/V \tag{2.8a}$$

This amounts to the adiabatic relationship $TV^{\gamma-1} = C$ notwithstanding accompanying heat transfer.

Substituting for P and Q in Equation 2.7 and inverting:

$$-Pd\varphi = \frac{d\Delta\tau}{\Delta\tau + Q(\varphi)/P(\varphi)}$$

The numerator is the differential coefficient of the denominator. With $\underline{P}$ and $\underline{Q}$ for respective mean values of $P(\varphi)$ and $Q(\varphi)$ over small interval $\Delta\varphi$:

$$\Delta\tau_{\varphi+\Delta\varphi} \approx -\underline{Q}/\underline{P} + (\Delta\tau_\varphi + \underline{Q}/\underline{P})e^{-\underline{P}\Delta\varphi} \tag{2.9}$$

2.4   Specimen solutions

Temperature and pressure solutions are functions of the four (dimensionless) parameters of the formulation: compression ratio r_v, specific heat ratio γ, temperature ratio N_T, and dimensionless heat transfer coefficient NTU.

$$T(\varphi)/T_C = f\{r_v, \gamma, N_T, NTU\} \tag{2.10}$$

Table 2.1 lists the data from which the dimensionless parameters are calculated. Holding S (linear size), h, and p_{ref} constant allows heat transfer effects to be thought of as a function of rpm (via ω – Equation 2.6).

Table 2.1 Values used in calculation of the dimensionless parameters of the solution

T_E	500 K
T_C	300 K
R	287 J/kgK
γ	1.4
h	50 W/m^{2}K
p_{ref}	1.0×10^5 Pa
D	50.0×10^{-3} m
S	25.0×10^{-3} m

Figure 2.4 plots instantaneous heat flux rate against crank angle *rpm* as parameter. The curve corresponding to lowest *rpm* (highest *NTU*) superimposes over the analytical solution for $T = \text{constant} = T_C$, that is, the 'isothermal' case. Successive doubling of *rpm* leads eventually to the adiabatic case.

This is an initial-value problem: initial conditions are $T = T_C$ at $V = V_1$. As volume reduces over the initial cycle *T* rises. For all but infinite *NTU*, average *T* exceeds enclosure temperature

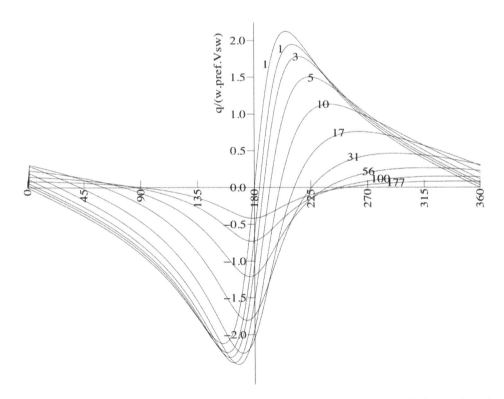

Figure 2.4 Heat flux rate as function of crank angle. Parameter is *rpm*. The symmetrical curve through (180, 0.0) is the analytical solution $dq/dt = pdV/dt$ for the isothermal case

T_C until cyclic equilibrium is achieved. Net cycle heat exchange is then equal to the (negative) area of the indicator diagram, $\int pdV$. For the data of Table 2.1 the largest negative value of $\int pdV$ occurs at 17 *rpm*. Traces in the figure are acquired at eventual cyclic equilibrium.

The asymmetry predicted from inspection of Equation 2.7 is apparent to varying degrees for *rpm* (i.e., for *NTU*) between isothermal and adiabatic extremes. Particularly noteworthy is that peak heat rejection rate (negative part of diagram) can exceed the 'isothermal' value!

2.5 'Realistic' Carnot cycle

'Realistic' refers to the gas process cycle, and not to the mechanical embodiment. For present purposes it means accepting that convective heat transfer takes time.

A reciprocating engine needs a crank or equivalent. The Scotch yoke generates the simple-harmonic motion previously supposed. Phases 2–3 and 4–1 remain adiabatic. Isothermal phases 1–2 and 3–4 of the ideal cycle are replaced by the relationship between T and V defined by Equation 2.9. The respective volume limits which they span need to be specified in terms of crank angle $\varphi = \omega t$. Equations 2.3 and 2.4 used earlier to establish transition volumes V_2 and V_4 no longer serve (except for $NTU = \infty$). The matter is dealt with by tracing compression phase 1–2 step-wise by numerical integration of Equation 2.9. The process proceeds until the evolving curve of T versus V intersects the adiabat passing through point (T_3, V_3). A similar incremental construction based on known conditions at 3 and 1 establishes point (T_4, V_4).

The indicator diagrams of Figure 2.5 were generated for the data of Table 2.1. The *NTU* value varies inversely as *rpm*, displayed as a parameter at the upper left. As might have been anticipated, cycle work, diagram area, and with it cycle work, decreases with *rpm*, that is, with decreasing time for heat transfer.

2.6 'Equivalent' polytropic index

The tradition of the engineering text-book would be to describe phases 1–2 and 3–4 in terms of a polytropic relationship $TV^{n-1} = C$, or $pV^n = C$. The present study affords this option – plus the benefit of a revealing insight.

If a numerical value of n properly defines the relationship between T and V over two states, 1 and 2, then it will define it at intermediate points. The proposition is readily checked as the process of step-wise integration of Equation 2.9 proceeds.

Between the current state and start point 1:

$$TV^{n-1} = T_1 V_1^{n-1}$$

Making index n explicit and using subscript c to distinguish compression values from those for polytropic expansion:

$$n_c = 1 + \frac{\log_e(T_1/T)}{\log_e(V/V_1)} \tag{2.11a}$$

Evaluating Equation 2.11 as integration proceeds reveals a strong variation of n_e with decreasing V. But for this finding it would be tempting to tabulate n_e and n_e against *NTU*, thus

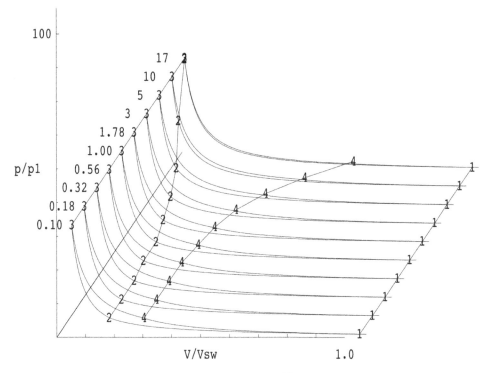

Figure 2.5 Indicator diagrams generated under the conditions of Table 2.1. Parameter is *rpm* (*NTU* inversely proportional to *rpm*)

streamlining the construction of *p-V* diagrams and the calculation of indicated work. In the circumstances, work is summed numerically as $\sum p \cdot dV$ during tracing of Equation 2.9.

The counterpart expression for n_e would be

$$n_e = 1 + \frac{\log_e(T_3/T)}{\log_e(V_g/V_3)} \tag{2.11b}$$

The mechanism for putting values to n_e and n_e might still be of interest, and so is included as Appendix A-2.

2.7 Réflexions

Symmetrical compression and expansion of a gas in a container at fixed temperature absorbs work, the amount per cycle depending, among others, on the intensity of convective heat exchange. After a sufficient number of cycles, heat rejected equals work input.

Despite the perfect symmetry of the volume variations, heat flux rates during compression are asymmetric with respect to those during expansion.

Use of a polytropic index *n* does not serve to relate *p* or *T* to *V* during the phases having limited heat transfer.

The *p-V* diagram of the 'static' Carnot cycle indicates positive work – but zero power (cycle speed zero). Matters can be made a degree more realistic on replacing the isotherms by heat exchange governed by coefficient of heat transfer h and instantaneous temperature difference, ΔT. This makes the cycle time-dependent – and thus speed-dependent.

With the above modification there is potential for power production. However, the instance explored suggests that the *rpm* range over which indicator diagram area remains positive is unpromisingly low.

A crank mechanism with désaxé offset e generates volume variations which are non-sinusoidal and which can be strongly asymmetric. One direction of rotation enhances the asymmetry of heat exchange. The opposite direction (or opposite sign of e) reduces it.

The thermal lag engine has only a single piston, and relies for the timing of the gas process events on the asymmetry of convective heat exchange. Thermal lag types driving crank mechanisms with désaxé offset are sensitive to the magnitude and sign of e.

3

What Carnot efficiency?

3.1 Epitaph to orthodoxy

In the normal course of events, this would be the place for a diagram showing a horizontal cylinder with opposed pistons in a succession of four positions. There would be an explanation of an 'ideal' Stirling cycle accompanied by claims of thermal efficiency equal to that of the Carnot cycle. Such an account may be found in almost every text and paper on the subject. The widely available coverage frees the present text to concentrate on a more critical look.

The concept attributed to Carnot is no more than that – a conceptual abstraction. The Stirling engine is a reality. The orthodox comparison is spurious: a substantial body of literature sees the commercial prospects of the Stirling engine in terms of its potential to set new efficiency levels. If your interest is based on this possibility, then save expense, time – and possibly your marriage – by pursuing instead a vision involving more predictable outlay and experimental development. The Stirling engine no more aspires to the Carnot efficiency than does the diesel engine or the steam turbine.

The limitation is consistent with the processes by which it functions: thermal conduction and forced convective heat transfer. These and the associated viscous flow phenomena are *irreversible*: they generate entropy. (Don't worry, *nobody* understands entropy: the concept of 'entropy creation' is useful for putting numbers to the performance penalty of irreversibilities. This is possible because the working formulae are of the cook-book variety – more user-friendly than the entropy concept *per se*.)

3.2 Putting Carnot to work

In the context of the Stirling engine a meaningful use for Carnot might take heat source temperature T_E to be that of the combustion reaction. Assuming the reaction to be more or less adiabatic, the figure is about 2000 °C (2273 K). The sink temperature T_0 must be ambient – say 20 °C, or 293 K. According to Carnot, heat entering at T_E is converted to work with an efficiency η_{Carnot} of $(2273 - 293)/2273 \approx 87\%$. The highest efficiencies reported for the Stirling engine are about 40% – that is, less than one-half of η_{Carnot}, and somewhat less than best figures for the diesel. For comparison, the expression for the efficiency of the 'ideal

Stirling Cycle Engines: Inner Workings and Design, First Edition. Allan J Organ.
© 2014 John Wiley & Sons, Ltd. Published 2014 by John Wiley & Sons, Ltd.

air-standard' Otto cycle is $\eta_{Otto} = 1 - 1/r_v^{\gamma-1}$, in which r_v is the compression ratio. (The formula for η_{Otto} should be used with caution: in Hitchcock's adaptation of *The Thirty-nine Steps* Mr Memory recites it on stage – and is shot!) Taking $r = 9$ (that is, 9:1) and $\gamma = 1.4$ gives $\eta_{Otto} = 0.58$ – or 58%. The brake thermal efficiency of the practical Otto-cycle engine (petrol) peaks at about 29% – half the ideal value. On this criterion the efficient, silent, burn-the-lot Stirling engine returns a worse showing than the petrol engine!

3.3   Mean cycle temperature difference, $\varepsilon_{Tx} = T - T_w$

Provisionally forget pumping losses and assume the inside wall surfaces of expansion and compression exchangers to be held at uniform $T_E = 900$ K and $T_C = 300$ K respectively. Assume also that the containing walls and regenerator are not subject to thermal short. This isolates the irreversibilities due to ε_{Txe} and ε_{Txc} at expansion and compression exchangers as the only cycle losses. These now account for the difference between indicated work per 'real' cycle and the indicated work suggested by the Schmidt analysis – or, indeed, by Carnot.

The Schmidt or Carnot efficiency η_{ideal} is:

$$\eta_{ideal} = (T_E - T_C)/T_E = (900 - 300)/900 = 0.66, \text{ or } 66\%$$

Obviously, no Stirling engine has ever returned an indicated thermal efficiency – let alone a brake efficiency approaching 66% – nor ever will.

At the expansion end, cycle mean value $\underline{\varepsilon}_{Txe} = \underline{T} - T_{we}$ must be negative to cause heat to flow *into* the working fluid. At the compression end $\underline{\varepsilon}_{Txc} = \underline{T} - T_{wc}$ and must take a positive numerical value.

To minimize the arithmetic, assume $\underline{\varepsilon}_{Txe}$ at the expansion end equal in magnitude to that at the compression end, viz: $\underline{\varepsilon}_{Txc} = -\underline{\varepsilon}_{Txe}$ (the two $\underline{\varepsilon}_{Tx}$ are of opposite sign). Re-writing the expression for indicated thermal efficiency with compensation for temperature 'error' ε_{Tx}:

$$\eta = (T_E + \underline{\varepsilon}_{Tx} - [T_C - \underline{\varepsilon}_{Tx}])/(T_E + \underline{\varepsilon}_{Tx})$$

With N_T for characteristic temperature ratio T_E/T_C:

$$\eta = (N_T + \underline{\varepsilon}_{Tx}/T_C - [1 - \underline{\varepsilon}_{Tx}/T_C])/(N_T + \underline{\varepsilon}_{Tx}/T_C)$$

The equation is linear in $\underline{\varepsilon}_{Tx}/T_C$, and is readily inverted to make $\underline{\varepsilon}_{Tx}/T_C$ explicit:

$$\underline{\varepsilon}_{Tx}/T_C = \{N_T(1 - \eta) - 1\}/(\eta - 2)$$

For $\eta = 40\%$ (= 0.4), and for the value of N_T given by the T_E and T_C introduced earlier:

$$\underline{\varepsilon}_{Tx}/T_C = \{3.0(1.0 - 0.4) - 1.0\}/(0.4 - 2.0) = 0.8/-1.6 = -0.5$$

$$\underline{\varepsilon}_{Tx} = -0.5T_C = -150\,°C$$

If 150 °C is some sort of cycle mean, imagine the peak value!

Several proprietary simulations have been fed the same values of T_E and T_C – with wildly varying results. One such item of dedicated design software – elaborate and not particularly cheap – gave $\underline{\varepsilon}_{Txe} = (-)\frac{1}{4}\,°C!!$

3.4 Net internal loss by inference

All engines (diesel, Stirling, gas turbine) operate with internal losses – heat transfer deficit, pumping loss, viscous dissipation, enthalpy flux, thermal conduction, 'adiabatic' loss, 'shuttle' heat transfer and mechanical friction. For specified heat input Q_{XE}, shaft work W is inevitably less than the maximum, W_{ideal} [J], of the relevant ideal cycle. The numerical difference between ideal work W_{ideal} and brake work W is equal to the sum of internal losses W_L [J]. If W_{ideal} can be calculated, the loss can be quantified.

Suppose that an engine tested in the laboratory at 35% brake thermal efficiency is delivering 350 W per kW of input energy rate. If the corresponding ideal (lossless) cycle for the same operating conditions has the 60% of our earlier calculation its 'output' would be 600 W per kW of input energy rate. The arithmetic suggests losses totalling 250 W, or 71% of net output.

In Chapter 4 of his 1991 text, Hargreaves presents test data on Philips' single-cylinder, rhombic-drive 30-15 engine – including overall thermal efficiency and flue losses, from which the brake thermal efficiency is readily back-calculated as a function of *rpm* and charge pressure. At peak power point (40 hp $\sim$ 30 kW), *rpm* are 2500, $T_E = 973$ [K], $T_C = 288$ [K], the brake thermal efficiency η is 32.5%. Mean expansion-end heat rate $Q'_{XE} = 30/\eta$ [kW] $= 92.3$ [kW]. Ideal (e.g., Schmidt-cycle) efficiency is 70%, and ideal work $\overline{W_{ideal}} = \eta_C \times 92.3 = 64.62$ [kW]. On this basis internal losses are $64.62 - 30.0 = \underline{34.62\ kW}$ – some 115% of rated output!

Figure 3.1 maps fractional loss W_L/W [–] for the 30-15 engine over the range of p_{ref} and *rpm* reported by Hargreaves. It shows the ratio becoming increasingly unfavourable at

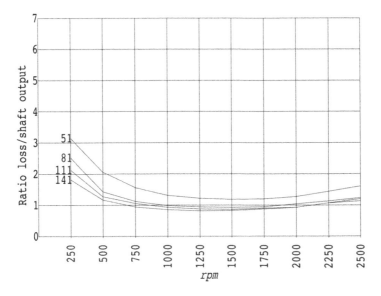

Figure 3.1 Net internal loss for Philips H_2-charged 30-15 engine as a function of *rpm*. The parameter is charge pressure p_{ref} (bar)

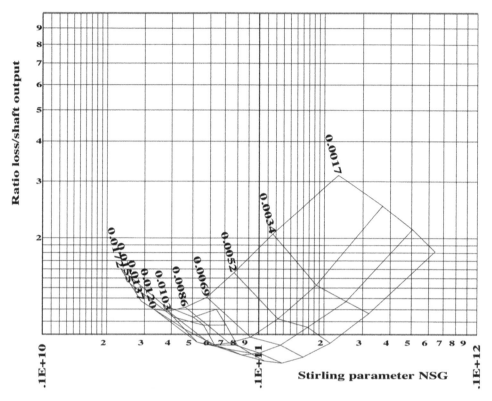

Figure 3.2   Figure 3.1 re-plotted for greater generality with net internal loss (fractional) as a function of Stirling number N_{SG}. The labelled parameter is Mach (or speed) parameter N_{MA}. The parameter of the four unlabelled curves is charge pressure p_{ref}

increasing distance from the peak-power point. Figure 3.2 re-maps Figure 3.1, in terms of the (dimensionless) parameters of Stirling engine operation.

If the efficiency of the ideal reference cycle is based on the temperature of the combustion gases, the picture of internal losses is even more dire.

If the preceding treatment seems a bit glib, how about a posher approach: *Availability Theory,* put to virtuoso use by Bejan (1994), and since applied with success to the regenerator, targets losses of all types in terms of *lost available work*[*]. In so doing it reduces each to a common denominator – whether heat transfer loss, flow friction or mechanical friction – an essential step, since the worst offenders are highlighted.

When heat flows at steady rate $\underline{Q}'_{XE}$ [W] across constant temperature difference ΔT_E, the rate of lost available work W'_{LXE} is $\underline{Q}'_{XE}|\Delta T_E/T_E|/N_T$, where N_T is the temperature ratio T_E/T_C [–]. Inverting:

$$\underline{\Delta T_E} = |W'_{LXE}/\underline{Q}'_{XE}|T_E/N_T \tag{3.1}$$

The four sets of dry, sliding polymer seals of the rhombic-drive machine (three of them high-Δp) suggest a mechanical efficiency of 75%. This would account for 25% of the 30 kW brake output, that is, for 7.5 kW of the 35 kW of internal losses. A small minority (2.5 kW) of the rest might be due to shuttle and conduction, leaving 25 kW of internal loss due to hydro-dynamic pumping and heat transfer deficit.

A number of studies suggest that, in a well-designed regenerator, the latter two losses are of comparable magnitude – and that this 50-50 balance is the target for the tubular exchangers. If the 25 kW remaining is arbitrarily divided (author's privilege) between the three exchangers, loss due to heat transfer deficit in the expansion exchanger is $\frac{1}{2} \times \frac{1}{3} \times 25 = 4.166$ kW. Substituting into Equation 3.1 suggests mean effective temperature difference $\overline{\Delta T_E} = 148\,°C$.

The figure results from values pulled out of thin air – but consider an hypothetical improvement to exchanger design which permitted ΔT_E to be halved without altering flow loss. The linearity of Equation 3.1 suggests a $\overline{\text{halving of}}$ loss due to heat transfer deficit. If this can be achieved without increasing flow loss, then $\frac{1}{2} \times 4.166$ kW $= 2.08$ kW is potentially added to brake output. Alternatively, the original 30 kW are available from reduced $\underline{Q'_{XE}}$ – compounding the reduction in ΔT_E.

Availability Theory has led independently to almost precisely the value of internal loss W_L resulting from the earlier back-of-envelope calculation!

3.5 Why no p-V diagram for the 'ideal' Stirling cycle?

Yet another old chestnut which will be conspicuous by its absence from these pages is the p-V diagram for the 'ideal Stirling cycle'. Appearances elsewhere are consistent with a widespread confusion between an *indicator diagram* of p versus V, and an *ideal cycle diagram* of p versus v, where V is instantaneous cylinder volume [m^3], and v, by contrast, is specific volume [m^3/kg]. In the case of the reciprocating internal combustion engine, the two diagrams look the same, respective areas (units of J and J/kg) differing only by a factor M, where M [kg] is the mass of working fluid taking part in the cycle. In the case of the internal combustion engine the academics get away with this because the properties of the working fluid are more or less uniform throughout the cylinder – particularly during compression and late expansion. This means that specific volume v and temperature T of a given particle are representative of the v and T of all particles at that instant during the cycle.

In the Stirling engine things could hardly be more different: the properties of a fluid particle occupying the expansion space at a given instant are quite distinct from those of a particle in the compression space at that same instant. There can be no unique ideal cycle diagram, neither of p versus v, nor of temperature *versus* entropy – of T versus s.

3.6 The way forward

The Stirling engine is widely lauded as aspiring to the Carnot efficiency. Aspire it might; achieve it does not and cannot. This is not to deny scope for improvement beyond efficiency levels so far demonstrated.

It appears essential to switch focus from the net work to the identification of individual loss components. Once these have been isolated and related to their respective causes, attention can turn to whittling them down.

4

Equivalence conditions for volume variations

4.1 Kinematic configuration

The cycle of interacting gas processes is driven by periodic variations of the 'live' volumes. The task of setting up a numerical model of the interactions therefore starts with a statement of volume variations as a function of crank angle φ.

Many crank mechanisms generate motions close to simple-harmonic. For the opposed piston, or 'alpha', configuration this allows simple expressions for $V_e(\varphi)$ and $V_c(\varphi)$. Employing the notation of Figure 4.1:

$$V_e(\varphi) = \tfrac{1}{2}V_E(1 + \cos\varphi) \tag{4.1a}$$

$$V_c(\varphi) = \tfrac{1}{2}\kappa V_E(1 + \cos[\varphi - \alpha]) \tag{4.1b}$$

In Equation 4.1b $\kappa = V_C/V_E$.

Angles α and β are the *kinematic* (i.e., crank) phase angles for the respective configuration – the angle by which the crank pin of the expansion piston (or displacer) leads that of the compression piston. In the case of the opposed-piston machine (only) α is also the *volume phase angle*, or *thermodynamic phase angle* – the angle by which the $V_e(\varphi)$ leads $V_c(\varphi)$. While Equation 4.1a for expansion volume $V_e(\varphi)$ extends to 'beta' and 'gamma' types, there is no analogous expression for $V_c(\varphi)$ because β is *not* the volume phase angle. Instead, $V_c(\varphi)$ must now be determined in terms of the *relative* locations of displacer and work piston.

Deriving $V_c(\varphi)$ in terms of β for the coaxial machine is not rocket science, but the 'obvious' approach, which proceeds via the algebra of the overlap of piston and displacer travel, is tedious. With characteristic flair, Finkelstein (1960a) short-circuited the problem, establishing by a process of inspection the conditions for the *equivalence* of $V_c(\varphi)$ between all three configurations. This has the benefit of making it possible to explore any one of the three variants as though it were of opposed-piston type.

For all configurations V_E denotes expansion volume *excursion*, $V_e(\varphi)_{max} - V_e(\varphi)_{min}$. Likewise, V_C denotes compression volume excursion, $V_c(\varphi)_{max} - V_c(\varphi)_{min}$. In the alpha machine

Stirling Cycle Engines: Inner Workings and Design, First Edition. Allan J Organ.
© 2014 John Wiley & Sons, Ltd. Published 2014 by John Wiley & Sons, Ltd.

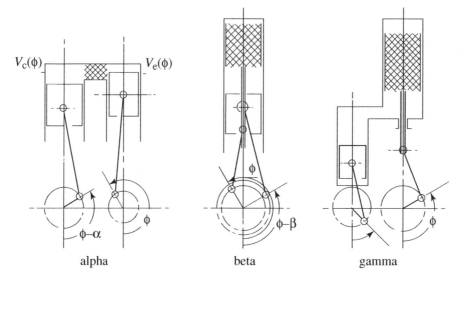

alpha   beta   gamma

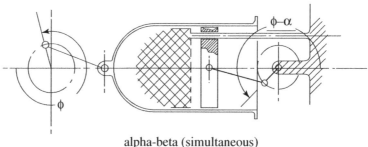

alpha-beta (simultaneous)

Figure 4.1 Schematic representation of the three principal configurations. After Finkelstein and Organ (2001), with permission of the American Society of Mechanical Engineers (ASME)

the expansion and compression piston respectively determine V_E and V_C. In beta and gamma types, displacer excursion (alone) determines V_E, while V_C is the result of *relative* motion between piston and displacer.

Ratio V_C/V_E is denoted κ and *defined* thermodynamic volume ratio *in all cases*. The kinematic displacement ratio λ is the ratio of the displacement (in isolation) over a stroke of the compression piston to that swept (in isolation) during a stroke of the displacer. In a coaxial machine with bores of uniform diameter, $\lambda \approx S_p/S_d$, where the S are respective strokes. (The approximation ($\approx$) reflects neglect of the cross-section area of displacer drive rod.)

$$\lambda = \sqrt{\{1 + \kappa^2 + 2\kappa \cos \alpha\}} \tag{4.2}$$

$$\beta = \operatorname{atan} \left\{ \frac{\kappa \sin \alpha}{\kappa \cos \alpha + 1} \right\} \tag{4.3}$$

$$\kappa = \sqrt{\{1 + \lambda^2 - 2\lambda \cos \beta\}} \tag{4.4}$$

$$\alpha = \operatorname{atan} \left\{ \frac{\lambda \sin \beta}{\lambda \cos \beta - 1} \right\} \tag{4.5}$$

If $\alpha <$ zero $\alpha = \alpha + \pi$.

Finkelstein constructed a chart for carrying out the conversions quickly. The original chart does not extend to cover values of λ recently found to be of particular interest ($\lambda <$ unity). A chart has therefore been generated from scratch as Figure 4.2a.

The values of λ and β established by Larque for easiest starting, best running and lowest-temperature operation of his parallel-bore, beta 'Vari-Engine' were 0.7 and 50 degrees approx-imately[1]. The value $\lambda = 0.7$ is located on the curved envelope at the upper left of Figure 4.2a, and traced along the curve of $\lambda =$ constant as far as the intersection with $\beta = 50$ degrees. This locates the point defining the κ and α giving, for the opposed-piston configuration, equivalent volume ratio and volume phase angle of the 'Vari-Engine'. Respective values are 0.7 and 135 degrees.

These are far from the 'traditional' values of unity and 90 degrees. Vaizey points out that a consequence of building $\kappa = 135$ degrees into an opposed-piston machine of V-configuration would be to forfeit the opportunity of optimum dynamic balancing.

The context of Finkelstein's formulation was the Schmidt analysis, but the algebra by itself embodies no thermodynamics: it is relevant to gas process models of any degree[2] of sophistication.

4.2 'Additional' dead space

Lambda machines differ from beta in that compression-space volume is always greater than zero by an amount given by Finkelstein's 'additional dead space' v_c, represented here by the ratio V_{dc}^+/V_E

$$V_{dc}^+/V_E = \frac{1}{2}(\lambda + 1 - \kappa) \tag{4.6}$$

Calculation of V_{dc}^+/V_E is applied only to beta and gamma types, so a value of λ will generally be to hand. However, a value of κ will not. Substituting from Equation 4.2:

$$V_{dc}^+/V_E = \frac{1}{2}\{1 + \lambda - \sqrt{(1 + \lambda^2 - 2\lambda \cos \beta)}\} \tag{4.6a}$$

There is no graphical representation of Equation 4.6a in the original 1960 treatment. For ease and speed of evaluation a nomogram is offered here as Figure 4.3. As with all such nomograms, the display affords instantaneous inversion, allowing V_{dc}^+/V_E to be read as $V_{dc}^+/V_E\{\lambda, \beta\}$, or λ as $\lambda\{V_{dc}^+/V_E, \beta\}$, or β as $\beta\{V_{dc}^+/V_E, \lambda\}$.

---

[1] Approximately: real mechanisms do not give simple-harmonic motion, and the phase between volume events varies somewhat around the cycle.

[2] Though not to the unsteady gas dynamics formulation, where the gas processes are driven by piston face *accelerations*.

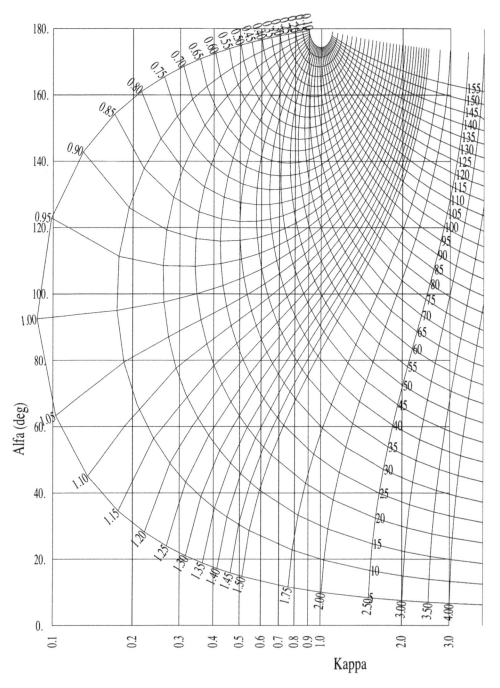

Figure 4.2a Graphical equivalent of Equations 4.2 and 4.3. Values of λ are marked against the left-hand curved boundary. The right-most column of (integer) parameter values elapsing vertically is β (in degrees)

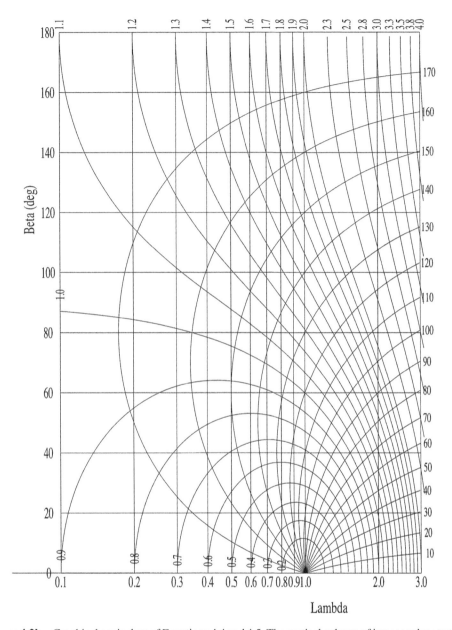

Figure 4.2b Graphical equivalent of Equations 4.4 and 4.5. The vertical column of integer values to the right are α (in degrees). The decimal parameter values label curves of constant κ. The figure is obviously an inversion of Figure 4.2a, reflecting the fact that Equation pair 4.4 and 4.5 is an inversion of Equation pair 4.2 and 4.3

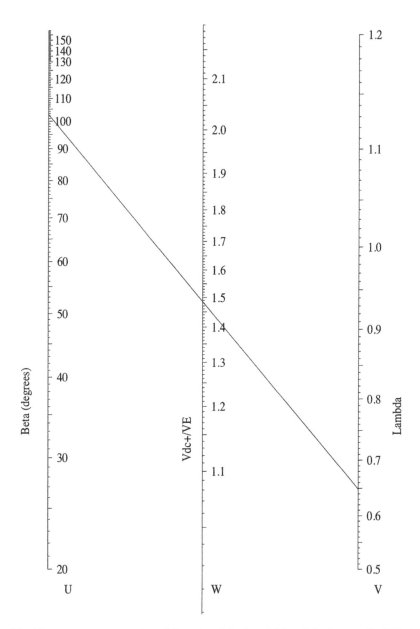

Figure 4.3 Nomogram representation of Equation 4.6a for additional dead space V_{dc}^{+}/V_E in terms of λ and β. The diagonal line is a sample solution for specimen parameter values $\lambda = 0.65$, $\beta = 102$ degrees

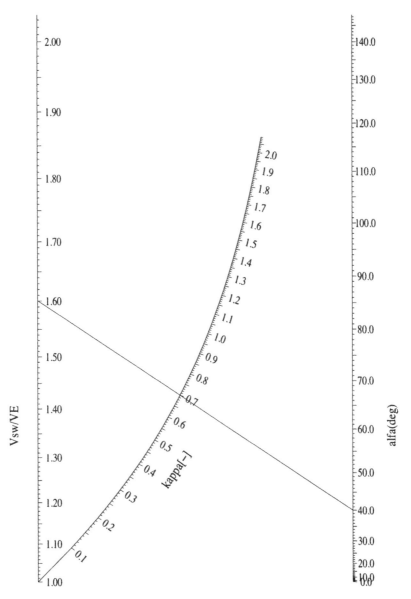

Figure 4.4 Nomogram representation of Equation 4.7 for ratio V_{sw}/V_E for opposed-piston (alpha) machine. Specimen solution (oblique line) is for $\kappa = V_C/V_E = 0.7$, volume phase angle $\alpha = 40$ degrees, giving $V_{sw} = 1.6\, V_E$

The recent experimental work by Larque, supported by analysis, has identified ranges of λ over which, depending upon temperature ratio N_T and internal *NTU* (Number of Transfer Units), specific cycle work *increases* with increasing dead space. On this basis a gamma engine might well perform more satisfactorily than an otherwise identical beta.

An alternative interpretation of Equation 4.6a is as the fractional amount of overlap of S_p and S_d in the uniform-bore beta machine.

4.3   Net swept volume

In beta and gamma machines net swept volume is V_{sw} is volume swept in a full stroke of the compression piston (with due allowance for cross-sectional area of displacer drive rod). Objective performance comparison calls for values of V_{sw} for all. (V_{sw} is the base of the indicator or *p-V* diagram.)

For the opposed piston (alpha) type:

$$V_{sw}/V_E = \sqrt{\{1 + \kappa^2 + 2\kappa \cos \alpha\}} \tag{4.7}$$

$$= \lambda \tag{4.7a}$$

With a value of V_E decided upon, Figure 4.4 serves for quick calculation of V_{sw} for two-piston machines.

For the original purpose of quantifying the dead-space 'handicap' of the lambda embodiment, Equation 4.6a is of limited interest, the more so following recent realization that increasing dead-space is not always a performance disadvantage – and sometimes the opposite. Its value is now seen to lie in allowing the simple expression for the volume variations of the gamma type to serve both machines *and handling the beta option by merely subtracting the numerical value of $V_{dc}{}^+$ from the compression-end dead space of the equivalent gamma.*

A gas-process model does not 'know' whether events in the gas path are occurring in an opposed-piston configuration (alpha) or in the displacer (beta or gamma) arrangement.

If expansion and compression space volumes of different engines were to follow identical respective cycle histories, then a simulation of one could amount to a simulation of all – regardless of mechanical layout. The lowermost diagram of Figure 4.1 is the equivalence condition reduced to a 'physical' counterpart.

In the two-piston, alpha machine kinematic phase angle is equal to volume phase angle – the angular difference at the crank between maximum (or minimum) expansion space volume and maximum (or minimum) volume of the compression space.

5

The optimum versus optimization

5.1 An engine from Turkey rocks the boat

The account by Karabulut et al. (2009) comes as something of a wake-up call: A single-cylinder, beta-configuration engine of 296 cm^3 swept volume operating between source temperature T_E of 200 °C and sink T_C of 27 °C (temperature ratio N_T of 1.577) achieved specific brake cycle work (Beale number N_B) of 0.0688. Relying on the temperature-ratio correction of Section 1.2.1 suggests that N_B at at the more common operating condition of $N_T \approx 3.0$ would be 0.125.

For an all-metal machine of 'hot-air' engine configuration this is no small achievement. But it is surpassed by the no-load characteristic: start-up was achieved at T_E of 93 °C. On removal of the heat source, no-load running continued down to $T_E = 75$ °C.

Here is a prototype which evidently wants to run – and to do so between un-promising temperature limits – in stark contrast to the majority of prototypes which manifestly prefer not to run – even when roasted to red heat. The evident free-running is the more remarkable for having been achieved with displacer driven by a mechanism not known for low friction – a slider similar to the 'rapid return' of a machine-shop shaping machine (Figure 5.1).

In general, intermediate temperature operation promises benefits: the range of available 'fuels' widens to include waste heat and geothermal sources. Operation from solar energy can be achieved despite relatively imprecise focusing and tracking. The requirement for sophisticated materials and fabrication techniques for hot-end components is relaxed.

It would be useful to be able to decipher the Karabulut account for the secret of the success.

The technical specification (Table 1 of the original account) yields nothing remarkable: Displacer L/D ratio is a healthy 3.5. There are no piston rings – and thus no ring friction, but the drive mechanism does not minimize side-loads. The stroke ratio $\lambda = S_p/S_d$ is 0.759. In Figure 4 of the original account, the volume maxima occur 110 degrees apart. Volume minima are separated by 138 degrees. If 'average' volume phase angle α has any meaning it is 124 degrees. From the same diagram, the thermodynamic volume ratio $\kappa = V_C/V_E = 1.066$.

Measured performance under load follows expectation: peak power and peak torque both increase initially with increasing charge pressure p_{ref}, and then fall off as NTU reduce in sympathy with further increase in p_{ref}.

Stirling Cycle Engines: Inner Workings and Design, First Edition. Allan J Organ.
© 2014 John Wiley & Sons, Ltd. Published 2014 by John Wiley & Sons, Ltd.

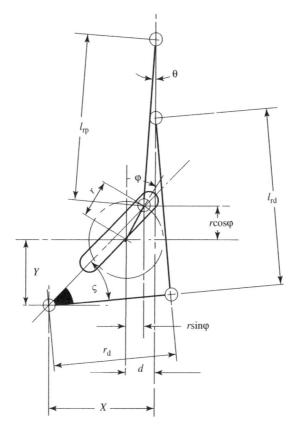

Figure 5.1   Skeleton of drive mechanism of Karabulut engine. *Not to scale!* – the figure serves to establish notation

Had volume variations derived from simple-harmonic motion (SHM), then values of λ, α, and κ would inter-convert (e.g., via the Finkelstein chart re-constructed here as Figure 4.2a), resulting in a unique value of kinematic phase angle β. However, the horizontal line through $\alpha = 124$ degrees fails to intersect the curve of $\lambda = 0.759$ at *any* value of κ. This is consistent with substantial deviation from SHM.

To this point the only distinguishing features of the design are: (a) values of α substantially in excess of the 'default' 90 degrees, (b) displacer stroke 32% greater than that of the piston, and (c) volume variations showing substantial digression from SHM. Further enquiry is necessary.

5.2   ... and an engine from Duxford

Like Karabulut et al., Vaizey envisages useful applications for the intermediate-temperature type. Intuition which has served him well in related endeavours (a thermal lag engine achieving 3100 no-load *rpm*) suggested offsetting the effect of decreasing temperature ratio by displacing an increasing fraction of gas between the temperature extremes. In the beta (coaxial) and gamma configurations this is achieved by a decrease in λ ($= S_{\mathrm{p}}/S_{\mathrm{d}}$).

Figure 5.2 Drive mechanism of the Vari-Engine. Reproduced by permission of Ian Larque

Vaizey collaborates closely with Larque whose 'Vari-Engine' was under construction at the time. The Vari-Engine was introduced in Section 1.2.6 as a small but 'serious' coaxial engine having slotted heat exchangers (as per the Philips MP1002CA) and annular regenerator formed by winding a bandage of dimpled foil.

Figure 5.2 shows the drive mechanism, consisting of twin parallel crankshafts rotating in a common sense and synchronized by a central idler gear. Piston and displacer are each driven off a radius arm pivoted remote from the cylinder axis. This arrangement minimizes side loads and associated friction. The hollow, cylindrical piston rod runs in a guide in the crank-case below the attachment points of the radius arms. The displacer rod is guided within the piston rod at a point also below the pin attachment point. The arrangement virtually eliminates side loads on piston and displacer themselves.

Variation of strokes S_p and S_d (and thus of λ) is achieved by use of interchangeable gears drilled for different crank-pin offset. Kinematic phase angle is altered by withdrawing the idler gear and advancing a crankshaft by one or more teeth. The process alters the vertical separation between piston and displacer, which is then corrected by adjusting the attachment point of one or both of the respective drive rods.

Comprehensive bench tests plotted brake power *versus rpm* for combinations of stroke ratio λ and kinematic phase angle β. Probably because the engine did not regain the exact cyclic equilibrium attained before being stopped for adjustment, the power curves are ragged – but sufficiently consistent to establish $\lambda \approx 0.75$ and $\beta \approx 50$ degrees as being the optimum combination for the engine under test. These were also the values at which no-load operation continued down to lowest T_E – surprising, perhaps, in view of the continuously changing temperature ratio N_T. The writer has witnessed no-load running with T_E of approximately 100 °C, followed by run-on lasting approximately 5 min following complete removal of the heat source. Vaizey has timed a similar event (run-down after heat removal) at 11 min.

Larque's optima translate to the opposed-piston 'alfa' configuration as $\alpha = 132$ degrees with $\kappa = 0.78$. The literature carries a number of independent optimization studies, including those of Finkelstein (1960a), Walker (1962), Kirkley (1962), Takase (1972), and the present author (Organ 1990). There is some disparity when results are plotted to a common format, but there is no single account which arrives at an optimum α even approaching 132 deg.

Conversely, when the Vari-Engine is set up with λ and β corresponding to the theoretical optima of α and κ, it is reluctant to run at all.

Academic optimization amounts to numerical perturbation of the Schmidt analysis. The latter is based on simple-harmonic volume variations and paints a 'static' picture of gas process events (i.e., is independent of cycle speed). The latter assumption is consistent with gas occupying the variable-volume spaces being uniformly at the nominal temperature of that space – T_C or T_E.

If reality and cycle model disagree, then the problem lies with the latter: either one (or more) of the founding assumptions is faulty, or its predictions are being incorrectly interpreted. Fortunately the Schmidt analysis itself can be used to diagnose the problem.

5.3 Schmidt on Schmidt

Setting the value $\lambda_{opt} \approx 1.5$ from published 'optimization' charts results in Vaizey and Larque engines becoming difficult to start. When eventually running, modest reduction in hot-end temperature brings running to a halt. Changing $S_p/S_d = \lambda$ to 0.75, that is, to almost the inverse of the theoretical 'optimum', transforms performance, leading to ready starting, increased no-load speed, and sustained no-load operation at dramatically-reduced T_E. In other words, performance takes on the characteristics of the engine of Kalibulut et al., where λ based on stroke ratio[1] S_p/S_d was 0.759.

The experimental result is, by definition the 'correct' one, meaning that the theoretical counterpart – the 'optimization' graph deriving from the Schmidt algebra – is defective. Two considerations overlooked in constructing the theoretical charts are:

Adiabatic temperature swing: this component of the compression and expansion phases depresses measured performance by adding to the load on the heat exchangers. Modifying Schmidt – as has been attempted (Organ, 1992) by assuming the variable-volume spaces to be adiabatic – fails to improve matters because keeping the algebra manageable means retaining the Schmidt assumptions as to gas temperature in exchangers and regenerator.

---

[1] The numerical value differs slightly from true displacement ratio to the extent of cross-sectional area of the displacer drive rod.

Mechanical friction: Dipl.-Ing. Peter Feulner, resident Stirling engine authority at Ricardo GmbH, cautions that dealing with friction in any general way is fraught with difficulty given the applicable range of bearings, seals, candidate mechanisms, and lubricants.

This, however, is not the point here: **If indicated work (p-V diagram area) is achieved at the cost of avoidably high pressure swing, then friction losses are unnecessarily high.**

And why the bold italics? Pressure swing is closely related to (volumetric) compression ratio r_v. By the time the relationship has been explored in Chapter 7 – A *question of adiabaticity* – it will have demonstrated potential to transform thinking about first-principles design.

If this assessment is correct, a way of comparing the r_v of different configurations is going to help steer design.

5.3.1 Volumetric compression ratio r_v

Finkelstein's re-working (1960a) of the Schmidt algebra really comes into its own at this point – and this means the trigonometry of the volume variations as opposed to the idealized thermodynamics. Compression ratio r_v is purely a geometric matter, and takes the same numerical value as does pressure ratio for the special case of $T_E = T_C$, that is, for $T_E =$ unity.

Pressure ratio is in terms of Finkelstein's 'compound parameter', ρ:

$$r_v = \frac{1 + \rho}{1 - \rho} \tag{5.1}$$

$$\rho = \frac{\sqrt{[(1/N_T)^2 + \kappa^2 + 2\kappa\cos\alpha/N_T]}}{1/N_T + \kappa + \nu} \tag{5.2}$$

With N_T set to unity:

$$\rho = \frac{\sqrt{[1 + \kappa^2 + 2\kappa\cos\alpha]}}{1 + \kappa + \nu} \tag{5.2a}$$

$$\nu = 2\Sigma V_{di}/V_E$$

Figure 5.3 displays r_v against volume ratio κ with volume phase angle α (in degrees) as parameter. The minimum of r_v occurs at almost precisely the κ and α which convert to the optima of λ and β established experimentally by Larque. While cyclic pressure swing and volumetric compression ratio are not the same thing they are closely related. The indicator (p-V) diagrams of Kalibulut's and of Larque's engines are almost certainly short and squat. The Schmidt algebra sheds further light.

5.3.2 Indicator diagram shape

An invaluable feature of Finkelstein's re-worked Schmidt algebra is that it lends itself readily to exchange of parameters κ and α (opposed-piston configuration) for λ and β (coaxial) – and vice versa. Figure 5.4 is a superposition of ideal indicator diagrams from the ideal, 'static' cycle as λ is increased from 0.5 to 3.0. The enclosed area (indicated work) increases from a low

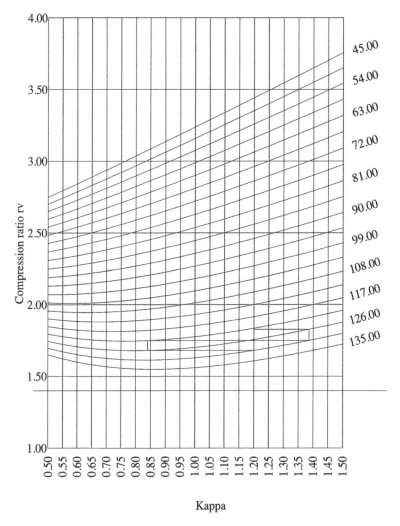

Figure 5.3 Graphical equivalent of Equation 5.1 – volumetric compression ratio r_v as a function of volume ratio κ with volume phase angle α as parameter

value for the inner, egg-shaped loop ($\lambda = 0.5$) via a maximum in the vicinity of $\lambda = 1.75$ – the optimum using Schmidt. This is accompanied by continuous increase in p_{max}/p_{min}. The area then decreases as λ increases beyond 1.75. The pressure swing p_{max}/p_{min}, however, continues to increase – at an escalating rate.

Figure 5.5 is an alternative depiction of the information content of Figure 5.4: the heavy curve shows ideal, specific cycle work[2] W_{sch}/pV_{env} passing through an optimum corresponding to the parameter values chosen; viz, $N_T = \overline{2.0}$ and v (Finkelstein's definition) $= 0.5$.

---

[2] V_{env} stands for *envelope volume* – the net volume swept in expansion and compression spaces. For the opposed-piston machine this is $V_E + V_C = V_E(1 + \kappa)$. For the gamma machine it is $V_E(1 + \lambda)$. Net volume swept in the compression

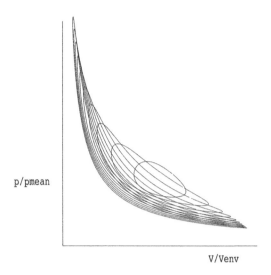

Figure 5.4 Superposition of ideal indicator diagrams while λ is varied from 0.5 (innermost, egg-shaped loop) to 3.0. (tallest, slimmest loop). The diagram area (indicated cycle work) first increases as λ increases towards the Schmidt optimum (approx. 1.75), then decreases. Pressure swing, increases at increasing rate

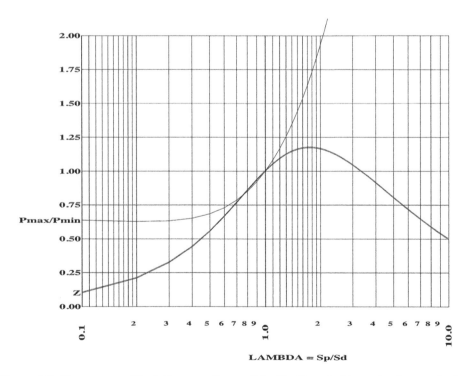

Figure 5.5 Heavy curve: W_{cyc}/pV_{env} according to Schmidt. $N_T = 2$, kinematic phase angle $\beta = 60$ degrees, dead-space parameter $\bar{\nu} = 0.5$. Light curve: pressure ratio $p_{max}/\underline{p}_{min}$. Both curves pro-rated by respective values at $\lambda = S_p/S_d =$ unity

By the value of λ which, in the absence of friction and the adiabatic effect would be λ_{opt}, pressure ratio has increased by 250%. This promises to more than offset the modest increase in ideal indicated work over the same interval.

5.3.3   More from the re-worked Schmidt analysis

Finkelstein's expression for instantaneous pressure p as a function of crank angle φ can be expressed in terms of p_{max}:

$$p = p_{max} \frac{1 - \rho}{1 + \cos(\varphi - \theta)} \tag{5.3}$$

$$\rho = \frac{\sqrt{[(1/N_T)^2 + \kappa^2 + 2\kappa\cos\alpha/N_T]}}{1/N_T + \kappa + \nu} \tag{5.4}$$

$$\theta = \text{atan} \frac{\kappa\sin\alpha}{1/N_T + \kappa\cos\alpha} \tag{5.5}$$

All terms are symmetrical in κ and N_T: pairs of values of κ and N_T which keep the product κN_T invariant yield a common value of p.

Dividing by κ the numerator and denominator of expressions for θ and ρ:

$$\rho = \frac{\sqrt{[1 + (1/\kappa N_T)^2 + 2\cos\alpha/\kappa N_T]}}{1 + 1/(\kappa N_T) + \nu/\kappa} \tag{5.4a}$$

$$\theta = \text{atan} \frac{\sin\alpha}{1/(\kappa N_T) + \cos\alpha} \tag{5.5a}$$

$$\nu = 2\Sigma V_{di}/T_i V_E$$

Expressing dead-space parameter as ν/κ gives the impression of a variable value. In fact, all that has changed is that the normalizing volume is now V_C rather than V_E – a more intuitive choice anyway.

From having been a function of four variables – N_T, κ, α, and ν – cycle pressure variation is now a function of three only – of $N_T\kappa$, α, and ν/κ. Suppose a pressure swing has been noted when $N_T = 3.0$ and $\kappa = $ unity. Now imagine N_T reduced to 1.5 with no change in phase angle α or dead space ratio ν/κ. Calculated pressure swing will remain unaltered provided the numerical value of κ is doubled.

---

space of the parallel-bore coaxial machine takes account of overlap of piston and displacer strokes: $V_E(1 + \kappa_{equiv})$, where κ_{equiv} is evaluated by conversion via Equations 4.4 and 4.5. V_{env} comes into its own because, in the context of optimization, use of the more 'obvious' choice of normalizing volume V_{sw} (base-length of the p-V diagram) leads to a perverse result. Mechanized optimization is mindless: it cannot distinguish an increase in a numerator (W) from a decrease in the denominator (V_{sw}).

Unlike the expression for pressure swing, that for specific cycle work is not fully symmetrical in κ and N_T:

$$Z_{\underline{p}Venv} = \frac{W}{\underline{p}V_E(1+\kappa)} = \frac{\pi}{1/\kappa+1} = \frac{(1/\kappa - 1/\kappa N_T)\sin\alpha}{(1/\kappa N_T + 1 + \nu/\kappa)}f_{\underline{p}}(\rho) \tag{5.6}$$

All other things being equal, specific work will inevitably reduce with the proposed halving of N_T. Pressure ratio remaining unchanged, a slimmer indicator diagram is inevitable. Both unwanted effects – mechanical friction and adiabatic heating component – are both enhanced relative to useful indicated work remaining.

The message of Figures 5.4 and 5.5 is more than corroborated. A design guide is emerging: if κ is *not* to be increased to compensate the decrease in N_T, then it must be left constant or be reduced. This tallies with $\kappa = 0.78$ which emerged from Larque's experiments.

The experiments of Vaizey and Larque may be thought of as achieving the *maximum numerical difference*, at prevailing operating conditions, between indicated cycle work and corresponding internal losses. Those losses include mechanical friction and an element of adiabatic temperature change.

The algebra of the theoretical search for the optimum cannot be modified so as reliably to reflect the realities of mechanical friction. On the other hand, one of the conditions for *minimum* friction is self-evident: *minimum loads on moving surfaces in contact*. A step in this direction – regardless of the detail of the mechanical embodiment – is **achievement of minimum cyclic pressure swing per unit of indicated cycle work**.

The Larque engine demonstrates smooth running down to low *rpm*. This is consistent with the kinetic energy in a slowly rotating flywheel readily overcoming the compression stroke – in other words, a low pressure ratio.

5.4 Crank-slider mechanism again

The drive mechanism of the Karabulut engine merits further study. Figure 5.6 shows the mechanism modified somewhat by incorporation of variable offset z' of slider track from crank-pin.

Applying Pythagoras' theorem gives piston gudgeon pin location y_p above the crank-shaft centre line:

$$y_p = r\cos\varphi + \sqrt{\{l_{rp}^2 - (\sin\varphi - d)^2\}} \tag{5.7}$$

Angle δ made by the slider with the horizontal is:

$$\delta = \text{atan}\{(\cos\varphi + Y)/(X - d + \sin\varphi)\}$$

The inclination of rod r_d to the horizontal is $\delta - \varsigma$. A design aim would be to keep rod l_{rd} as nearly vertical as possible, so that instantaneous height y_d of the gudgeon pin driving the displacer is given approximately by:

$$yd = -Y + r_d\sin(\delta - \varsigma) + l_{rd} \tag{5.8}$$

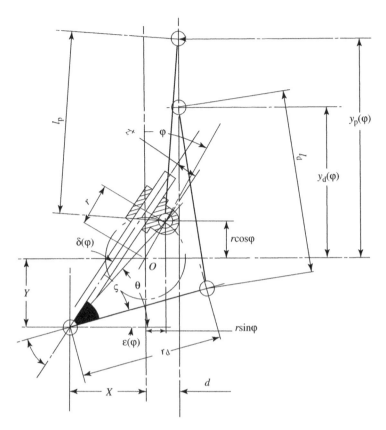

Figure 5.6   Karabulut crank-slider mechanism modified by inclusion of variable offset z' between slider track and crank-pin

Volume variations follow upon cycling the calculation from $\varphi = 0$ to $\varphi = 2\pi$ and noting the value of φ at which the difference $y_p - y_d$ is a maximum. This is the point at which, with zero clearance between piston crown and bottom of displacer, the two members would *just* come into contact.

The maximum of y_d (V_e minimum) occurs at the value of φ for which crank throw r is perpendicular to the axis of the slider. The difference $y_p - y_d$ is found by inspection, so φ corresponding to maximum y_d can be noted during that process.

5.5   Implications for engine design in general

This chapter has made heavy use of the Schmidt algebra to probe matters which that very algebra excludes from consideration – an approach with which it would be easy to find fault.

The volumetric compression ratio r_v, on the other hand, is purely a matter of geometry, independent of any assumptions about gas processes. The best performance of the Larque engine was achieved at *almost precisely* the λ and β giving the minimum value of r_v on Figure 5.3.

The study has nothing to offer the symmetrical Rinia engine (four-cylinder alpha): the first harmonic of α is 90 degrees whether one likes it or not, while κ is precisely $1 - (d/D)^2$, where D is cylinder bore and d is piston drive-rod diameter – again whether one likes it or not.

With mechanical friction acknowledged, beta (coaxial) and gamma types are fundamentally different from alpha (opposed piston). Both pistons of the latter are loaded by the instantaneous difference Δp in pressure between gas path and buffer space. In beta and gamma types only the work piston is so loaded: the displacer rod 'feels' the same Δp – but its small diameter makes the effect insignificant. The major diameter of the displacer feels only the pumping pressure drop. On this basis, the load on its drive mechanism is limited largely to inertia (and to friction itself) and so can be designed accordingly: small loads can be carried by small bearings, and small, lightly loaded bearings connote low friction loss.

Efforts[3] abound to develop drive mechanisms achieving the volume variations of the 'ideal' Stirling cycle. If implemented they would yield pressure ratios in excess of those due to simple-harmonic motion. High accelerations would compound already increased bearing loads. The adiabatic component of temperature swing would be increased.

The 'ideal' cycle has been imposed retrospectively by academics in their well-meaning search of a common denominator for the study of power cycles. It has, however, nothing whatsoever to do with the realities of a functioning engine.

A lower pressure swing for a given indicated work suggests a lower peak pressure. For a given creep-to-rupture stress the possibility arises of lighter hot-end components. There are convincing reasons why future optimization studies might move from p_{mean} as reference pressure in favour of p_{max} – despite the greater convenience of the former.

---

[3]E.g., that posted as *http://www.youtube.com/watch?v=YvtekHPzbUw*

6

Steady-flow heat transfer correlations

6.1 Turbulent – or turbulent?

Attempts to represent the functioning Stirling engine in a numerical model span half a century. Five decades might be considered sufficient for the discipline to have reached maturity. Un-resolved misgivings centre on[1] the tubular exchanger duct and on widespread use of engineering design data of questionable relevance – steady-flow *correlations* between Stanton number *St* and Reynolds number *Re* and between friction factor *Cf* and *Re*. These depict a range of *Re* ($0 < Re < 2300$) for which flow is laminar, and a transition range above the upper limit of which flow is turbulent. Table 6.1 contrasts this with realities of conditions in the Stirling engine.

Characteristic time between peak and 'zero' flow in the exchanger is one quarter cycle. At 1500 *rpm* ¼ cycle takes 0.01 s. Depending upon operating conditions and hydraulic radius, this is less than the estimated half-life (Section 6.2 following) of an eddy – and thus insufficient time for turbulence to decay. If there is insufficient time for decay, what are the mechanisms of formation and persistence?

Chances of resolving the matter can be weighed against expert opinion quoted here from the authoritative account by Moin and Kim for Scientific American (see URL in the References):

> For a phenomenon that is literally ubiquitous, remarkably little of a quantitative nature is known about it. Richard Feynman, the great Nobel Prize-winning physicist, called turbulence "the most important unsolved problem of classical physics." Its difficulty was wittily expressed in 1932 by the British physicist Horace Lamb who, in an address to the British Association for the Advancement of Science, reportedly said, "I am an old man now, and when I die and go to heaven there are two matters on which I hope for enlightenment. One is quantum electrodynamics, and the other is the turbulent motion of fluids. And about the former I am rather optimistic."

---

[1] And extends to the annular gap of the 'hot-air' type. (Wire-mesh regenerator subject to a wide range of different concerns – addressed under separate heading.)

Stirling Cycle Engines: Inner Workings and Design, First Edition. Allan J Organ.
© 2014 John Wiley & Sons, Ltd. Published 2014 by John Wiley & Sons, Ltd.

Table 6.1 Contrast between conditions of Stirling engine operation and those under which 'design' correlations acquired

Correlations	Stirling engine
Straight duct; flow axi-symmetric; no secondary flows	*Duct not straight; flow not axi-symmetric; secondary flows*
Pressure constant	$p_{max}/p_{min} \approx 3$
Flow uni-directional, steady, incompressible	*Flow oscillating, unsteady and compressible*
(For Cf-Re correlation) flow isothermal	*Flow not isothermal*
Flow free of entry effects and fully established	*Significant entry effects; flow not fully established*
Re is criterion of flow regime (laminar versus turbulent) –	*Numerical value of Re is no guide to flow regime (laminar/turbulent). Instead –*
0 < Re < 2300: laminar	*flow accelerating: laminar*
Re > 2300: turbulent	*flow decelerating: turbulent*
Regenerator:	
Acquired for single or small number of screens with nominal temperature gradient	*Large number of screens (500–1000) under intense temperature gradient (15 000 K/m)*

It is possible that Feynman and Lamb had in mind 'free' turbulence, as observed in wakes and plumes. This sets the 'forced', periodic turbulence of the Stirling engine somewhat apart. An authority having first-hand experience is John Corey, editor of *Stirling Machine World*. His response to the concern about lack of time for eddy decay is:

"... the vortex at the heart of an eddy has energy storage (as well as angular momentum), and so the time-dependence of its behavior can be viewed in an inertial manner. Just as a child's swing, when periodically excited (IF at the right frequency) can store the energy of those inputs and establish a persistent state of motion; so the vortex, periodically excited by the flow and pressure wave of the cycle, can develop and sustain persistent turbulence, even though the development takes many cycles of excitation.

This inertial perspective clarified for me some very interesting data we took decades ago, that I did not really understand until this recent exercise. We had dynamic pressure sensors in some ducts, driven by oscillating flows, and we looked for the onset of turbulence in sub-cycle time. At first (above frequencies where flow was fully laminar and the pressure traces cleanly sinusoidal), the onset would appear in the latter half of each cycle, as pressure began to diminish, and while the intensity of pressure disturbances grew with increasing frequency, the onset point within a cycle did not move to an earlier point, as might be expected; but rather the tail, extended over the zero-crossing and began to infect the rising part; until in the end, all was chaotic (save for the fundamental)! I now think that your point about lacking time to establish an eddy is the reason for non-advancing initiation, and the inertial persistence the reason for the sustain (musically speaking) that eventually closed the choral loop of turbulence."

A comprehensive report by Simon and Seume reproduces (at Figure 3-7 of the original) traces recorded experimentally by Hino et al. These are consistent with John Corey's account and furnish dramatic graphical illustration.

6.2 Eddy dispersion time

Rules of thumb exist for eddy dispersion time. Derivations are harder to come by. Prof. Rex Britter suggests decay time should be proportional to $l/\underline{u}$, where l [m] is characteristic eddy dimension and $\underline{u}$ [m/s] is speed. The 'back of envelope' approach which follows leads to the estimate of 'half-life' $\Delta t_{1/2} = r_h Re/\underline{u}$. With hydraulic radius r_h [m] for characteristic linear dimension, this is in line with Britter's suggestion.

To acquire a feel for numbers it is assumed that the half-life of an eddy is of the same order as half-life for collapse of the laminar velocity profile on instantaneous removal of the driving pressure difference. In Figure 6.1, which shows an element of cylindrical duct:

$$\tau_w = -\mu \partial u/\partial r \tag{6.1a}$$

$$= \tfrac{1}{2}\varrho \underline{u}^2 Cf \tag{6.1b}$$

In the laminar regime $Cf = 16/Re$, for which the RHS reduces to $2\underline{u}\mu/r_h$. Driving force due to pressure difference Δp over duct length dx equates to the product of wall shear stress τ_w with the peripheral area over which it acts:

$$\Delta p A_{ff} = -2\underline{u}\mu p_w \mathrm{d}x/r_h$$

Sudden removal of Δp is equivalent to imposition of a decelerating force balanced by the product of mass element $\varrho A_{ff}\mathrm{d}x$ with its deceleration d$\underline{u}$/dt:

$$\mathrm{d}\underline{u}/\underline{u} = -\left(2\mu/\varrho r_h^2\right)\mathrm{d}t \tag{6.2}$$

Integrating:

$$\log_e(\underline{u}_1/\underline{u}_0) = -\left(2\mu/\varrho r_h^2\right)(t_1 - t_0)$$

Time $\Delta t_{1/2}$ for $\underline{u}$ to fall from $\underline{u}_0$ to half of that value is given by inverting Equation 6.3:

$$\Delta t_{1/2} = -\left(\varrho r_h^2/2\mu\right)\log_e \tfrac{1}{2} = -(r_h/8\underline{u})Re \cdot \log_e \tfrac{1}{2} \tag{6.3}$$

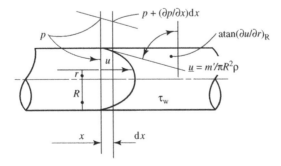

Figure 6.1 Notation for algebra of time of decay of laminar velocity profile based on instantaneous removal of driving pressure difference

Hydraulic radius r_h of the exchanger slots of the Philips MP1002CA air engine is 0.2 mm. Assuming air at 12 bar and 30 °C ($\mu = 0.017 \times 10^{-3}$ Pas) flowing steadily in a slot gives $\Delta t_{1/2} = 0.0098$ s ≈ 0.01 s.

This is approximately 0.01 s, which at the 1500 *rpm* of the MP1002CA is precisely the time for ¼ cycle, that is, the time from maximum or minimum flow rate to zero.

6.3 Contribution from 'inverse modelling'

An alternative to conventional cycle simulation is *'inverse modelling'*. The algebra is formulated in such a way as to make gas path geometry is explicit. This enables lengths L_x, hydraulic radii r_{hx} and numbers of ducts n_{Tx}, together with regenerator specification to be read from charts corresponding to the designer's choice of operating conditions (p_{ref}, *rpm*, working gas) and output power requirement.

A feature is separation of per-cycle exchanger losses – loss due to imperfect heat transfer, and loss due to hydrodynamic pumping. At the level of the algebra, inverse modelling has been demonstrated as workable. On the other hand, algebra and flow data are independent.

Attempts at validation have involved applying to each of the 'benchmark' engines in turn – GPU-3, MP1002CA, 400 hp/cyl, V-160, and so on, with the unexpected finding that back-calculated heat transfer loss component exceeds by two orders of magnitude that due to pumping. Expansion and compression exchangers both exhibit the disparity – although regenerator losses remain in balance in every case. This outcome – if it reflects reality – suggests an opportunity lost at the original design stage: the opportunity to enhance heat transfer capacity for a small sacrifice in pumping power. In other words, when back-calculated from steady-flow correlations, exchanger design is way off-balance in each and every engine.

The unlikely scenario demanded a second, independent formulation. The outcome – two orders of magnitude imbalance – remained. Two further alternative 'inverse models' have since been formulated – both fully-functional. All four algorithms had been coded independently. All four yield the same outcome – the same gross imbalance.

The sole common feature of all four formulations are the steady-flow heat transfer and flow correlations – $StPr^{2/3}$ – Re and Cf – Re. Conditions in the Stirling engine not being conducive to laminar flow, the turbulent correlation is used to cover the full Reynolds number range:

$$St = 0.023Re^{-0.2}Pr^{-0.4} \tag{6.4}$$

Reynolds' analogy for the tubular duct in steady flow is:

$$Cf \approx 2St \tag{6.5}$$

Respective velocity profiles (Figure 6.2) insist that, at given mass flow rate m', wall shear stress $\tau_w = \mu \partial u/\partial y$ is higher in turbulent flow than in laminar. However, St and Cf calculated via Equations 6.4 and 6.5 for Re below the magic 2300 are *lower* than laminar values at the same Re. The 'inverse model' integrates over the entire cycle. It thus enters the range ($0 < Re < 2300$), under-estimating pumping loss and over-estimating heat transfer loss (under-estimating St) – consistent with the (ostensible) exchanger imbalance already referred to.

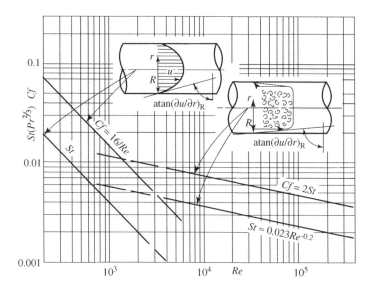

Figure 6.2 Laminar and turbulent correlations for steady, fully established, isothermal flow in the rectilinear, cylindrical duct. Use of the algebraic expression for the turbulent correlation at Re below about 2×10^3 predicts St and Cf values *lower* than those for laminar flow. This conflicts with the generally held picture of velocity gradient in the boundary layer

The picture which emerges is that, while flow can be neither laminar nor fully developed, the steady-flow alternative – the turbulent correlation (Equation 6.4) – is not appropriate as it stands to the Re range covered.

Applicability of standard correlations to the Stirling context has long been open to question. Misgivings are now compounded:

- Shear stress at the wall $\tau_w = -\mu \partial u / \partial y$ at given instantaneous Re cannot conceivably be that of the steady-flow correlation at the same Re.
- 'Flush ratio' now connotes more than just the possibility of an un-flushed plug oscillating pointlessly within the 'isothermal' exchanger: any such plug now has turbulence history fundamentally different from that of fluid entering afresh at either end once per cycle.
- If, by chance, at peak cycle Re, turbulence in the plug achieves intensity comparable to that at the same Re in steady flow, then it must have been via a mechanism different from the steady-flow case.
- Impressive accuracy of performance prediction is claimed for simulations based on 'piece-wise steady-state' use of traditional St-Re and Cf-Re correlations. Is the Stirling engine, after all, so perverse that any old data input leads to a high-fidelity performance prediction?
- As far as this book is concerned, those scaling options which embody the steady-flow correlations (via values of the exponent of Equation 6.4) now require re-evaluation. Exceptions are Dynamic Similarity and *FastTrack* (Chapters 9 and 12, respectively), both of which bypass empirical correlations.

If turbulence does, indeed, prevail over the cycle, and if Equations 6.4 and 6.5 nevertheless do not serve, then the question arises as to what functional relationship should be resorted to for numerical modelling of the exchange processes.

Equation 6.4 was of the form $St(Pr^c) = aRe^b$. Assuming that a similar expression applies – *but with coefficients a and b altered to a' and b'*, then the material of the following Section leads to a numerical value b' and paves the way for the value for a'.

6.4 Contribution from Scaling

In summary so far:

- At rated operating *rpm* conditions for establishment of laminar flow do not exist.
- If flow cannot be laminar then it has to be turbulent.
- Use of the *Cf-Re* and *St-Re* correlation appropriate to steady flow in the turbulent regime (Re > 2300) lead to substantial over-estimate of cycle heat transfer deficit, and corresponding under-estimate of concomitant pumping loss. The discrepancy is consistent with under-estimates of *St* and of *Cf* from Reynolds correlations $St = 0.023Re^{-0.2}Pr^{-0.4}$ and $Cf \approx 2St$ (Reynolds' analogy).

First-principles thermodynamic design (other than by CFD) relies on availability of workable correlations. The challenge of acquiring these experimentally under conditions representative of the Stirling engine gas processes is daunting, possibly explaining why there are no advertised plans to tackle it.

Engineering and politics have one thing in common: both have been described as being '*the art of the possible*'. This justifies pursuing alternatives to direct experimentation. Subject to certain assumptions, this is possible through *inverse scaling* from measured performance of 'benchmark' engines – P-40, GPU-3, V-160, and so on.

Imagine a computer simulation based on appropriately formulated numerical solution of the one-dimensional conservation equations, incorporating correlations of the form $St(Pr^c) = aRe^b$ and acknowledging Reynolds' analogy (*Cf* = 2*St* in turbulent pipe flow). The simulation is now applied to a range of engines substantially similar in terms of temperature ratio, dead-space distribution, phase angle and volume ratio under conditions where respective numerical values of specific cycle work N_B are comparable, that is, at rated operating conditions – speed *rpm* and charge pressure p_{ref}. With L_x and r_h for flow-passage length and hydraulic radius respectively, the algebra of the formulation insists that numerical values of $(L_x/r_h)^{1+b}/N_{SG}{}^b$ (where Stirling parameter $N_{SG} = p_{ref}/\omega\mu_{ref}$) should be approximately equal to a (dimensionless) constant, say, C_{sim}. Evidence for the constancy is offered in Chapter 6 of the author's 1997 text,

Now suppose that, under the turbulent conditions of the Stirling exchanger, *St* correlates to *Re* such that $St = a'Re^{b'}f(Pr)$, where a' and b' are constants, but of different respective numerical value from those of the original *a* and *b* of the steady-flow correlation. Suppose in addition that $(L_x/r_h)^{1+b'}/N_{SG}{}^{b'}$, calculated for a range of *real* engines, yields near-constant value C_{exp}, then the value of b' may be inferred from a plot of L_x/r_h versus N_{SG}.

Taking logs:

$$(1 + b')ln(L_x/r_h) - b'ln(N_{SG}) \approx lnC_{exp}$$

Dividing through by b':

$$[(1 + b')/b']ln(L_x/r_h) - ln(N_{SG}) \approx lnC_{exp}/b'$$

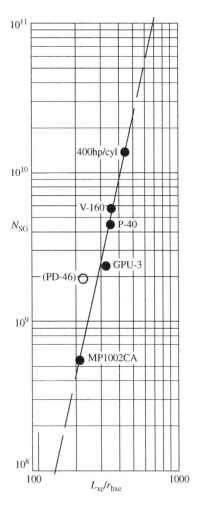

Figure 6.3   Stirling number N_{SG} versus length ratio L_{xe}/r_{hxe} for six 'benchmark' engines. The gradient yields the value of b' (exponent in $St - Re$ correlation)

This is a linear relationship of the form:

$$f(N_{SG}) \approx m \cdot f(L_x/r_h) + C'$$

in which m is the gradient of the curve – measured from the plot. The 'experimental' value of b' follows from $(1 + b')/b' = m$.

Figure 6.3 is a log-log plot of N_{SG} versus L_{xe}/r_{hxe} (expansion exchangers) for the six well-documented engines. Gradient $m = \partial\log_{10}(N_{SG})/\partial\log_{10}(L_x/r_h) = 4.3883$, giving $b' = -0.2951$ versus $b = -0.2$ of the steady-flow correlation for the cylindrical duct.

The investigation leaves the value of a' temporarily un-resolved. Subject, however, to the original assumption as to the algebraic form of the correlation for periodic, turbulent flow

$(St(Pr^c) = a'Re^{b'})$ a numerical value has been established for the (negative) gradient of the curve of log(St) versus log(Re). Being steeper than that of the steady-flow case, the line through any verified combination of St and Re within the turbulent regime now intersects the steady-flow laminar curve at a lower value of Re than previously. While the feature is not definitive, things have moved reassuringly in a direction consistent with turbulence initiating (and decaying) at a value of Re lower than for the steady-flow case.

The reader who has managed, against all the odds, to remain awake to this point may be asking why the corresponding correlation of the 1997 text (i.e., Figure 6.9) failed to yield this information. It was the belief at the time that the traditional steady-flow correlation for the turbulent regime was appropriate. Similarity parameter L_{xe}/r_{hxe} and N_{SG} were accordingly plotted[2] with respective exponents $1 + b$ and b – that is, with 1.2 and 0.2 respectively. Scale factors for the two log axes being equal, the straight line appropriate to thinking at the time lay at 45 degrees, as then drawn. The best fit, on the other hand, is a line of steeper gradient. A re-plot with the value of b' now proposed, viz., as $(L_{xe}/r_{hxe})^{1.2951}$ versus $N_{SG}{}^{0.2951}$ is now convincingly fitted by a line at 45 degrees gradient.

6.5   What turbulence level?

The flow conditions of interest de-couple Reynolds number from turbulence intensity, that is, Stanton number St and friction factor Cf are no longer known functions of Re. On the other hand, with due account of the distinction between u [function of radial coordinate r, viz, $u(r)$] and $\underline{u}$ (mean axial velocity), Reynolds number Re, friction factor Cf and wall shear stress remain defined.

On the evidence of the previous Section, rated performance of developed engines is consistent with St of the multi-tube exchanger relating to Re as $St(Pr^{\frac{2}{3}}) = a'Re^{-0.2951} \approx a'Re^{-0.3}$ where the numerical value for parameter a' remains to be determined.

If further investigation upholds the finding, all that is required to determine parameter a' is a value for local, instantaneous pressure gradient dp/dx corresponding to an arbitrary value of Re at turbulence intensity representative of engine operation.

Assume an experiment in which flow is made to oscillate under representative conditions. And suppose that dp/dx recorded at some characteristic Re (e.g., mean value $\underline{Re}$) exceeds the steady-flow turbulent correlation value at $\underline{Re}$ by factor x. Provided the excess owes little to inertia (i.e., reflects the changed turbulence intensity), wall shear stress and friction factor are also x times the counterpart steady-flow values, viz, $Cf_{\underline{Re}} = 2 \times 0.023(x\underline{Re})^{-0.2}$, where factor 2 reflects Reynolds' analogy linking St and Cf.

But according to the previous Section, the relationship between Cf and Re under the turbulent flow conditions of the functioning engine is $Cf_{\underline{Re}} = 2 \times a'Re^{-0.3}$. The two expressions coincide at $Re = \underline{Re}$, allowing a' to be isolated:

$$a' = 0.023x^{-0.2}\underline{Re}^{0.1} \tag{6.6}$$

---

[2]NB with x and y coordinate directions interchanged.

Depending upon the information available from experiment, acquisition of a value for a' can be approached differently: Setting Equation 6.1a equal to Equation 6.1b:

$$\tfrac{1}{4}CfRe = \partial(u/\underline{u})/\partial(r/r_0) \tag{6.7}$$

Equation 6.7 is merely a relationship between the algebraic definitions of Cf and Re: no particular flow regime or correlation is implied. Multiplying the right-hand side by $-\mu\underline{u}/r_0$ (numerical values all known) yields local wall shear stress, τ_w.

7

A question of adiabaticity

7.1 Data

Theoretical study has the capability of providing support at two levels: (a) exploration of the performance envelope of the *genre* and (b) minimization of trial-and-error in achieving the potential of a candidate design.

Even the most humble cycle description invokes data[1]: the 'textbook' cycle, the Schmidt and other 'isothermal' models tacitly assume values of zero and infinity respectively for coefficient of dynamic viscosity μ [Pas] and heat transfer coefficient h [W/m²K]. In total contrast, the various 'adiabatic' analyses are based on h = zero [W/m²K] for the variable-volume spaces.

Neither value qualifies as fact, but a reality which does is that indicator (p-V) diagrams from published 'adiabatic' treatments – for example, that of Urieli and Berchowitz (1984, p. 87), differ little from the Schmidt when both are computed for common values of the respective parameters – temperature ratio, volume ratio, volume phase angle, and dead-space ratio(s). Either (a) it does not matter which of the two extreme values is chosen for h or (b) extant formulations of the 'adiabatic' model do not fully reflect the implications of h = zero.

If either extreme were correct then the thermal lag engine could not function, so reality must lie with some intermediate value. A simple bench test sheds light out of all proportion to the experimental overhead.

7.2 The Archibald test

Figure 7.1 indicates a sealed cylinder, some 50 mm (2 inches) diameter and 75 mm (3 inches) long, containing air and attached to a simple Bourdon-tube pressure gauge. With container and contents pre-cooled to zero deg. C, pressure is atmospheric and the gauge indicates zero. The container is heated to, say, 400 °C, plunged into a water/ice mix and the time for air temperature T to return to the start value noted, as indicated by the fall in pressure.

Since John Archibald notified the experiment by e-mail, the author has described it on numerous occasions and challenged hearers to estimate temperature decay time. Estimates

---

[1] *Things known or assumed as facts, and made the basis of reasoning or calculation.* Shorter Oxford Dictionary, 1993.

Stirling Cycle Engines: Inner Workings and Design, First Edition. Allan J Organ.
© 2014 John Wiley & Sons, Ltd. Published 2014 by John Wiley & Sons, Ltd.

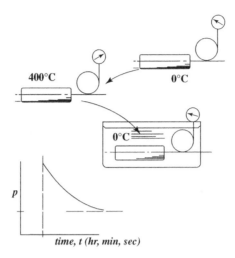

Figure 7.1 The Archibald Test – schematic representation

have ranged from hours via minutes to a few seconds. Readers might care to make a guess before turning the page for the value* timed from the YouTube demonstration. (The author himself would have guessed at a matter of minutes.)

The volume of the air cylinder (some 145 cm^3) is representative of the displacement of many Stirling engines. On the other hand, cycle time at the 1500 *rpm* of Philips' MP1002CA air-charged engine is 0.04 s, or about one hundredth of the decay time estimated from the Archibald test. This short interval occupies the steepest part of the decay curve which, in turn, applies to a gas which is nominally still.

The Archibald test will be recognized as an improvized demonstration of Newton's law of cooling, the analytical expression for which is in terms of heat transfer coefficient, h, which can now be estimated.

7.3   A contribution from Newton

A fixed volume V contains mass $m = \rho V$ of fluid at uniform temperature T in contact with a surface of area A_w initially also T. At time t_0 surface temperature changes suddenly to T_w and remains at that value. The rate of response of T can be expressed in terms of h:

$$hA_w(T_w - T) = \rho c_v V dT/dt \tag{7.1}$$

T_w being constant, nothing is changed by re-expressing in terms of instantaneous temperature difference ΔT, where $\Delta T = T_w - T$:

$$d\Delta T/dt = -(hA_w/\rho c_v V)d\Delta T/dt$$

The history of ΔT as time t elapses is:

$$\Delta T = \Delta T_0 \exp(-hA_w/\rho c_v V)(t - t_0) \tag{7.2}$$

Table 7.1 Specimen heat transfer coefficients h. From Schneider (1955)

Free convection of air across pipes	1.135	W/m^2K
Liquid metal in pipes	36 900	W/m^2K
Drop-wise condensation – steam on plates	113 560	W/m^2K

For a given fluid contained in a known volume, a run of the Archibald test provides a specimen numerical value for both ΔT and $t - t_0$, and thus a spot value of h.

Inverting Equation 7.2 to expose h:

$$h = \frac{-\log_e(\Delta T/\Delta T_0)}{t - t_0} \rho c_v V/A_w \qquad (7.3)$$

The fluid is air, and a test starts with the contents of V at atmospheric pressure and zero degrees C. Density ρ is therefore about 1.275 $[kg/m^3]$. Being equal to m/V the value of ρ does not change as pressure and temperature vary during the experiment. Specific heat at constant volume $c_v = 0.72 \times 10^3$ $[J/kgK]$. Values for V and A_w follow from diameter (50 mm) and length (75 mm) of the metal container. Assuming that time taken for ΔT to drop by 90% of ΔT_0 is the time over which pressure dropped by the same percentage of start value gives $h \approx 25$ W/m^2K.

Table 7.1 lists specimen values converted from the imperial units (BTU/hr $\cdot$ ft$^2 \cdot$ °F) of the account by Schneider (1955).

The 'obvious' question arises: is h of 25 W/m^2K a low value (near-adiabatic) or high (near-isothermal)? The question is irrelevant, because a value of h in isolation – even of h in conjunction with a value for temperature difference – does not determine the thermal response of the fluid. The latter depends on a ratio: (*capacity of the wetted surface to transfer heat*)/(*capacity of the enclosed fluid for thermal take-up*). The numerical value of the ratio is unaltered if both capacities are expressed as *per unit time*.

The concept can be seen in action by setting up an energy balance for a simple cylinder of variable-volume with walls at uniform temperature T_w.

7.4 Variable-volume space

Figure 7.2 represents a cylinder closed at its upper end and sealed from below by a piston. There are no ports, so total mass of gas processed, M, is invariant. The piston is driven at uniform angular speed by a 'real' mechanism, that is, motion is not confined to being simple-harmonic.

The piston stroke is s, and axial clearance at inner dead centre position is c, so that compression ratio r_c is $1 + s/c$. Allowing for horizontal offset (désaxé) e between axes of crankshaft and cylinder will permit eventual examination of asymmetry of expansion and compression processes.

Over time increment dt:

$$dQ - pdV = dU$$

Adapting Equation 7.1:

$$hA_w(T_w - T) - pdV = Mc_v dT/dt \qquad (7.5)$$

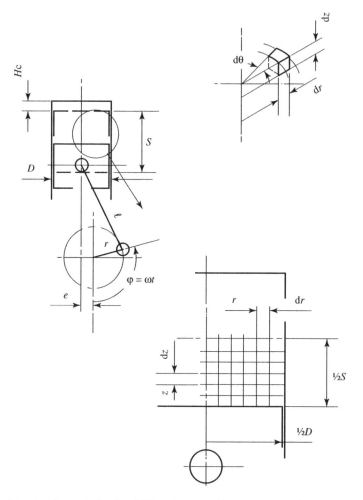

Figure 7.2   Notation for analysis of variable-volume working space enclosing fixed mass of gas M

Replacing p by MRT/V (ideal gas law), using $R/c_v = \gamma - 1$, expressing T as $T - T_w + T_w$, abbreviating $(T - T_w)/T_w$ to $\Delta\tau$ and re-arranging:

$$d\Delta\tau/d\varphi + \Delta\tau(\gamma - 1)\{V^{-1}dV/d\varphi + \eta A_w/A_0\} = -(\gamma - 1)\tau_w V^{-1}dV/d\varphi \qquad (7.6)$$

In Equation 7.6 crank angle $\varphi = \omega t$, so that ωdt has become $d\varphi$. Symbol $\eta = hT_w A_0/(\omega p_0 V_0)$. With $T_w/p_0 V_0$ eliminated with the aid of the ideal gas equation, and making the substitution $c_p/R = \gamma/(\gamma - 1)$, symbol η emerges as Finkelstein's 'dimensionless heat transfer coefficient' $hA/(\omega M c_p)$.

The equation is of the form $d\Delta\tau/d\varphi + \Delta\tau P(\varphi) = -Q(\varphi)$, which can be re-written:

$$-P d\varphi = \frac{d\Delta\tau}{\Delta\tau + Q(\varphi)/P(\varphi)}$$

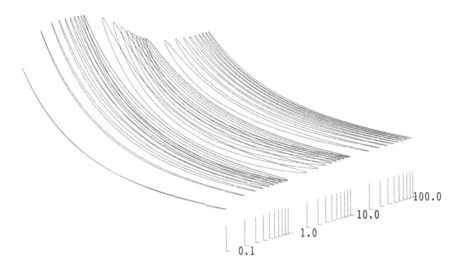

Figure 7.3 p-V loops corresponding to work lost per cycle for $0.1 < \eta < 100$

The numerator is the differential coefficient of the denominator. With $\underline{P}$ and $\underline{Q}$ for respective mean values of $P(\varphi)$ and $Q(\varphi)$ over small interval $\Delta\varphi$:

$$\Delta\tau_{\varphi+\Delta\varphi} \approx -\underline{Q}/\underline{P} + (\Delta\tau_\varphi + \underline{Q}/\underline{P})e^{-\underline{P}\Delta\varphi} \tag{7.7}$$

A problems threatens when $\underline{P} \to 0$, but is readily dealt with by returning to Equation 7.6:

$$\Delta\tau_{\varphi+\Delta\varphi} - \Delta\tau_\varphi \approx \tau_w(\gamma - 1)dV/V \tag{7.8a}$$

which, with due account of the definition of P, predictably reflects the adiabatic relationship $TV^{\gamma-1} = $ constant. The parallel-cylinder configuration allows the term $V^{-1}dV/d\varphi$ to be evaluated as $-(ds/d\varphi)/(S + C - s)$.

Integrating Equation 7.7 over a cycle with η set to zero generates the curve $pV^\gamma = $ constant, which encloses zero area. Repeating with η set to infinity generates a different curve, $pV = $ constant. For values of η between the two extremes, heat is exchanged irreversibly between gas and cylinder, and a result is curves enclosing finite area. All curves are traced anti-clockwise, so indicated work $\int pdV$ is negative.

Figure 7.3 has been generated by integrating Equation 7.7 for the data of Table 7.2. The left-most trace is for $\eta = 0.1$ which evidently gives near-adiabatic processes. Negative work peaks in the range $1.0 < \eta < 10$. Increasing η by three orders of magnitude over the start value (to $\eta = 100$) fails to yield isothermal conditions (zero loop area).

Back calculating h from $\eta = 100$ gives the astronomically high $h = 7.23 \times 10^5$!

7.5   Désaxé

In Figure 7.2 the centre-line of the cylinder is off-set by horizontal distance e from that of the crank-shaft. With crankshaft angular velocity uniform, and positive in the sense indicated,

Table 7.2 Parameter values used in constructing Figure 7.3

cylinder diameter, d	20.0 mm
crank throw r (= semi-stroke)	12.5 mm
con. rod centres l	25.0 mm
désaxé offset e	+10.0 mm
compression ratio, r_c	3:1
pressure at V_{max}	10 atm (10^6 Pa)
wall temp, T_w	300 °C

peak angularity of the connecting-rod is greater on the compression (low pressure) stroke than on the expansion (high pressure) stroke. A result of practical consequence is a reduction in the peak piston side load relative to that due to the symmetrical layout. Moreover, the compression stroke occupies less than 180 degrees of crankshaft rotation, the expansion stroke occupying correspondingly more: rate of compression exceeds rate of expansion.

Reversing the sign of e or, alternatively, keeping e the same and reversing direction of crankshaft rotation at any given value of η leads to a change in calculated negative work.

7.6 Thermal diffusion – axi-symmetric case

It would be an elementary matter to apply the Archibald test to a cylinder of length several times the diameter. The transient thermal conduction phenomenon is then expressed in cylindrical polar coordinates by the equation:

$$\partial^2 T/\partial r^2 + r^{-1}\partial T/\partial r = \alpha^{-1}\partial T/\partial t \qquad (7.9)$$

In Equation 7.9 r is local radius, t is elapsed time. Thermal diffusivity α is defined as $k/c_p\rho$ and has units of [m²/s].

This being the sole parameter, all cases sharing a common numerical value of α have identical solution history: the temperature solution (record of T vs t) for given boundary and initial conditions (outer radius r_0 [m], initial temperature field and temperature disturbance at the periphery) *is a function of thermal diffusivity α alone*! For copper of 99% purity $\alpha = 0.11 \times 10^{-3}$ [m²/s]. For comparison, α for H_2 at ambient pressure and temperature is – wait for it – *the same*!! This means that the Archibald test carried out in container of pure copper amounts to a test of a rigid cylinder of H_2. (It is also a test of a solid copper cylinder of the same size.)

To preserve generality (and to take advantage of Dynamic Similarity), numerical solution best proceeds in terms of fractional radius r/r_0, fractional temperature T/T_{ref} and dimensionless time $\omega t = \varphi$. This introduces Fourier modulus $N_F = \alpha/\omega r_0^2$. Thermal diffusivity α for air at ambient pressure and temperature is approximately 10% of that for H_2. To achieve the same N_F between H_2- and air-charged engines at given ω (i.e., at given *rpm*), the numerical r_0^2 for the air engine must be one-tenth of that for H_2. In turn this calls for the ratio r_{0air}/r_{0H_2} to be $\sqrt{10} = 3.16$. If cycle history of gas displacement is to be nominally the same, then dead volume V_{dx} of the two tubular exchanger bundles must be the same. Dead volume $V_{dx} = n_T \pi r_0^2 L_x$, where n_T is number of tubes. For similarity of temperature history, the air-charged engine requires 10 times the number of (smaller) tubes. The result is consistent with results of scaling studies to be featured in later chapters.

7.7 Convection versus diffusion

The initial look at the Archibald test was in terms of heat transfer coefficient h. The diffusion equation (Equation 7.9) achieves a description of the entire location-temperature record with no mention of h. Is there incompatibility between the two accounts of transient response?

On the contrary, the link lies in the definition of h, which also has the more revealing name *film coefficient*. When heat flows by convection at rate q' from a solid surface at temperature T_w to a fluid of bulk temperature T_∞ it does so by conduction through a *thermal boundary layer*:

$$q' = -kA\partial(T - T_w)/\partial y|_{y=0}$$

But this has already been expressed as:

$$q' = hA(T_w - T_\infty)$$

Equating the two versions:

$$h/k = \frac{\partial(T_w - T)/\partial y|_{y=0}}{T_w - T_\infty}$$

The result will be more general for being dimensionless. This can be achieved by working in terms of fraction distance y/L where L is a length characteristic of local geometry. The obvious choice for internal flows is hydraulic radius r_h. The resulting group is Nusselt number N_{Nu}:

$$N_{Nu} = hr_h/k \qquad (7.10)$$

7.8 Bridging the gap

The evolving treatment remains short of representing conditions in the variable-volume spaces. Cardinal features of the latter are: (a) cyclic variation of volume – and pressure – and (b) motion of the gas relative to the cylinder surface.

A step in the right direction can be made with the aid of diagrams at the right-hand side of Figure 7.2. These show a slice through the axis of the vertical cylinder and an element of gas representative of conditions at radius r. All elements are subject to common instantaneous pressure p. The assumption of symmetry about the z axis means that, at any instant, temperature is uniform in the circumferential (θ) direction.

The unsteady energy equation for an element is:

$$q' - w' = u'$$

Net rate of energy accumulation due to thermal diffusion in the radial direction is:

$$q_r' = kr\Delta\theta\Delta z\{r(\partial^2 T/\partial r^2) + (\partial T/\partial r)\}\Delta r$$

In the z direction:

$$q_z' = kr\Delta\theta\Delta r(\partial^2 T/\partial r^2)\Delta z$$

For any given element, dm calculated from initial conditions p_0 and T_0 is invariant.

With $\underline{T}$ to denote instantaneous mean temperature of the element, rate of change of internal energy is:

$$d\underline{u}/dt = c_v dm\partial \underline{T}/dt$$

Elemental work rate assuming that the sides of the element remain orthogonal:

$$w' = pd(\Delta V)/dt = pd\{r\Delta\theta\Delta z\Delta r\}/dt$$

The symbolic description is prepared for numerical solution by re-casting in terms of finite differences. With j and k as labels for increments in r (radial) and z (axial) directions respectively:

$$\partial^2 T/\partial r^2|_{j,k} \approx \frac{T_{j+1,k} - 2T_{j,k} + T_{j-1,k}}{\Delta r^2}$$

$$\partial T/\partial t|_{j,k} \approx \left(T_{j,k}{}^{t+dt} - T_{j,k}{}^t\right)/\Delta t = \Delta T_{j,k}|_t/\Delta t$$

The dependent variables are the increment or decrement $\Delta T_{j,k}|_t$ in $T_{j,k}$ at mesh intersections, increments/decrements Δr, the Δz and the single (common) change in dp. Each is evaluated over time increment Δt.

At each station k of z the sum of all Δr_k is set to zero, viz, $\sum \Delta r_k = 0$. At each station j of r $\sum \Delta z_j$ is equal to the change in vertical location of the piston. At each station k along axis z a further constraint is available on noting that $\partial T_{0,k}/\partial r = zero \approx (T_{1,k} - T_{-1,k})/2\Delta r$.

Coefficients of the resulting set of linear equations form a square array which is 'solved' for $\Delta T_{j,k}|_t$, and so on by a software library routine such as SIMQX. Results are added to respective $T_{j,k}|_t$ to yield values $T_{j,k}|_{t+\Delta t}$, and so on corresponding to the end $t + \Delta t$ of the integration interval.

Figure 7.4 gives perspective views of specimen temperature solutions, both resulting from several cycles of compression and expansion. The outermost particles remain at cylinder surface temperature T_w, so the rim defines T_w in both examples.

The upper figure offers a fresh interpretation of 'adiabatic': a cylindrical core of gas little influenced by wall temperature and responding to pressure swing via $T/p^{(\gamma-1)/\gamma} = $ constant. Temperature disturbance is confined largely to the boundary film.

Inflating k by three orders of magnitude and examining a sequence of plots reveals a temperature wave propagating radially inwards. The lower diagram of Figure 7.4 is a 'still'

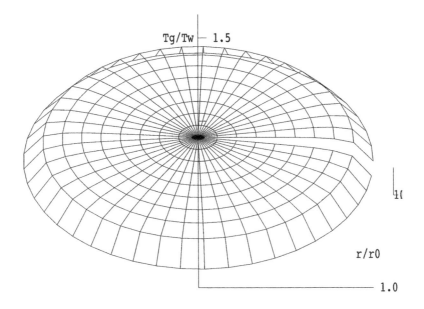

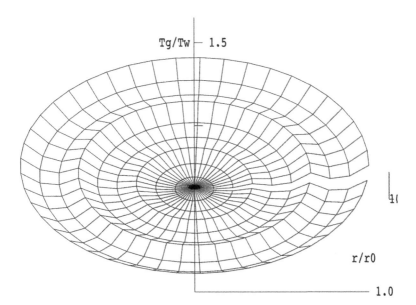

Figure 7.4 Specimen temperature surfaces at cylinder mid-height. Upper: air at conditions of Table 7.2 ($k = 0.0261$ W/mK). Lower: air with conductivity k artificially inflated by three orders of magnitude

from a representative sequence. Similarity with the diurnal wave phenomenon analysed by Schneider (1955) will be recognized.

7.9   Interim deductions

The performance potential of a Stirling engine is frequently discussed in terms of the thermal conductivity k of the working fluid – air versus He versus H_2. The numerical value of k – *in isolation* – reveals next to nothing.

It is not known at what point in history the gas cycle of the Stirling engine first became saddled with the notion of isothermal (and constant-volume) processes. Boyle's Law dates from 1662. Charles' Law was proposed considerably later in the 1780s, but remained obscure until Gay-Lussac's exposition of 1802.

It is hard to see how the notion of the 'isothermal' process could have enjoyed currency prior to awareness of the concept of *adiabatic*. Rankine's 1845 account of the cycle was argued graphically, and was in terms of *process paths*, including 'isodiabatic' paths. Such a path is not *necessarily* adiabatic, but the usage suggest awareness of 'adiabatic' as a limiting case.

Had the variable-volume processes been treated from the outset as Rankine's 'diabatic' – and other cycle phases evaluated accordingly – it is possible that the Stirling engine might have progressed differently towards commercial exploitation. A high price since has been paid for the burdensome – and inappropriate – association with isothermal phases, with the Carnot cycle and for the resulting epidemic of shattered illusions.

8

More adiabaticity

8.1 'Harmful' dead space

When it comes to Stirling engines, Theodor Finkelstein[1] got few things wrong – perhaps only one: In his *tour-de-force* '*Optimization of phase angle and volume ratio for Stirling engines*' (1960a), and in the course of introducing dead space parameter ν, he asserts:

'... and ν which denotes harmful dead space to be minimized by the designer.'

Two experimental observations, arrived at independently, highlight the fallacy.

A rhombic-drive engine, air-charged, and of 120 mm bore × 76 mm stroke, failed to turn a crank – pressurized or otherwise – despite manufacture to the highest machining standards by Vaizey. The remedy? Additional dead volume introduced to the compression space by a re-designed piston!

Bomford has built several large-displacement engines. Charged with air at atmospheric pressure, and with only the displacer annulus as regenerator, they have powered his skiff at 3–4 knots on the River Severn.

Expectation was therefore high when he ventured into slotted heat exchangers, pressurization and multiple, wire-wool packed regenerators. The engine failed to run – until removal of an annular sleeve *increased* the dead space of the expansion-end gas path substantially. The new willingness defied *increased* hydraulic radius and correspondingly *decreased* heat transfer capacity (numerical value of *NTU*)!

Both observations are consistent with the systematic enquiry subsequently launched by Vaizey, pursued independently by Larque and recounted in the previous chapter.

Under the 'isothermal' assumptions none of the above can happen. And there is worse: when the isothermal cycle model is used to find the 'optimum' specification (volume ratio κ, phase angle α) the result can be inferior to an arbitrary design.

---

[1]He was a pioneer of computer-dating in the days of punched cards, becoming something of a property tycoon. He eventually had a residence in the same Beverly Hills street as former US president Ronald Reagan. When funds ran low he profitably worked the stock exchange.

Stirling Cycle Engines: Inner Workings and Design, First Edition. Allan J Organ.
© 2014 John Wiley & Sons, Ltd. Published 2014 by John Wiley & Sons, Ltd.

Back-calculation from the 'Archibald test' (Chapter 7) leaves no doubt that conditions in the variable-volume spaces at realistic *rpm* are closer to adiabatic than to isothermal.

With Schmidt and 'isothermal' models discredited, the challenge is to find a replacement. Merely suppressing heat exchange in the variable-volume spaces, as attempted by the author (Organ 1992), by Urieli and Berchowitz (1984), and by Walker (1980), does not result in the observed sensitivity to dead space variation.

A full-blown simulation should respond realistically. On the other hand, the ideal is the simplest formulation which captures the phenomenon.

The criterion excludes formulations based on finite elements, on unsteady gas dynamics (Characteristics) and on kinetic theory (Chapter 21). Remaining options rely on a convective heat transfer 'model' in the form of correlations between Stanton number St and Reynolds number Re, viz, $StPr^{2/3} = f(Re)$. These reflect experimental conditions which are totally un-representative, namely, steady, incompressible, uni-directional flow in straight ducts free of secondary flows. A difficult choice arises between:

(a) Retaining the traditional gas path – out of phase volume variations linked by the gas path – and embodying heat transfer data alien to the context.
(b) Setting up an 'equivalent' steady-flow cycle in which the use of heat transfer correlations can be justified.

If option (b) appears far-fetched, it is worth considering that, for temperature ratio $N_T >$ unity, *any* picture of the Stirling cycle specified in a unique $T\text{-}s$ or $p\text{-}v_s$ trace implies a uni-directional, steady-flow sequence.

8.2 'Equivalent' steady-flow closed-cycle regenerative engine

Figure 8.1 shows schematically a gas path to explore what this alternative might have to offer. Compression and expansion are achieved aerodynamically (by turbo-machinery) or by positive displacement – provided the latter is steady and continuous. Either way, these processes are

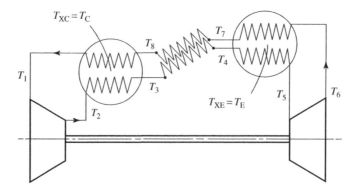

Figure 8.1   Steady flow 'equivalent' Stirling engine

assumed adiabatic – already closer to the reality of the Stirling engine than the traditional isothermal assumption. With r_p for pressure ratio:

$$T_2 = T_1 r_p^{(\gamma-1)/\gamma} \tag{8.1}$$

The compression exchanger has n_{Txc} tubes, each of internal diameter d_{xc} and length L_{xc}, as per that of the reciprocating engine except (a) that individual tubes are straight and (b) that the tube set is twinned – one for outflow, the other for return. Dead space doubled, but this is irrelevant to the steady-flow case. The twin exchangers operate at internal surface temperature T_{XC}. In terms of inlet temperature T_2 and Number of Transfer Units NTU_{xc} (uniform in steady flow):

$$T_3 - T_{XC} = (T_2 - T_{XC})e^{-NTUxc} \tag{8.2}$$

A symmetrical counter-flow recuperator substitutes the usual regenerator. Being a fictive component it can have the volume porosity $\P_v$ and hydraulic radius r_{hr} of the wire-mesh stack on both outward and return surfaces. Assuming fluid properties identical in both streams:

$$T_7 - T_4 = T_8 - T_3 \tag{8.3}$$

The tube bundle of the expansion exchanger has surface temperature T_{XE}. With gas inlet temperature T_4 and uniform Number of Transfer Units NTU_{XE}:

$$T_5 - T_{XE} = (T_4 - T_{XE})e^{-NTUxe} \tag{8.4}$$

Adiabatic expansion:

$$T_6 = T_5 \, r_p^{-(\gamma-1)/\gamma} \tag{8.5}$$

The expansion exchanger for the return pass 6–7 replicates that of forward pass 4–5, so has surface temperature T_{XE}:

$$T_7 - T_{XE} = (T_6 - T_{XE})e^{-NTUxe} \tag{8.6}$$

Under present assumptions, gas to gas temperature differences either side of the recuperator separating wall(s) are equal in magnitude. This makes wall temperature T_W equal at any internal location to the mean of the temperatures of hot and cold streams.

At the extremities:

$$T_{WE} = \tfrac{1}{2}(T_7 + T_4); \quad T_{WC} = \tfrac{1}{2}(T_8 + T_3) \qquad \text{(a)}$$

From the steady-flow solution for realistic NTU_r:

$$T_8 - T_{WC} \approx (T_{WE} - T_{WC})/NTU_r \qquad \text{(b)}$$

Substituting (a) into (b) and cancelling the common multiplier ½:

$$T_3(1 - 1/NTU_r) + T_4/NTU_r + T_7/NTU_r - T_8(1 + 1/NTU_r) = 0 \tag{8.7}$$

On the return pass 8–1 through compression exchanger:

$$T_1 - T_{XC} = (T_8 - T_{XC})e^{-NTU_{XC}} \tag{8.8}$$

Equations 8.1 to 8.8 are linear in the eight unknown temperatures. On allocating numerical values to T_{XC}, T_{XE}, and the three NTU values (section following) solution is routine.

8.3 'Equivalence'

At given temperature ratio N_T and dimensionless dead space ν, nominal compression ratio in the (reciprocating) Stirling engine may be expressed in terms of phase angle α and volume ratio κ – or, equivalently, in terms of the kinematic conversions, β and λ. If the steady-flow 'equivalent' Stirling engine follows the regenerative turbine in having an optimum compression ratio, it will be possible to back-calculate the α and κ (or β and λ) corresponding to the optimum.

Equivalence here is notional, but all that matters in exercises of this nature is consistency between different cases scrutinized.

Numbers selected for NTU_{XC}, NTU_r, and NTU_{XE} are based on rms values back-calculated from simulation of the reciprocating counterpart – 2.0, 100.0 and 1.8 respectively. Arbitrary values can be handled.

The phenomenon which it is hoped to explain is the experimentally-noted effect of varying stroke ratio λ ($\approx S_p/S_d$) in the parallel-bore 'beta' engine – particularly when operating at intermediate temperature ratios.

8.4 Simulated performance

For all diagrams nominal pressure ratio r_p is calculated assuming a constant value (57 degrees) of kinematic phase angle β but allowing kinematic displacement ratio λ and dead space parameter ν to vary. Heat rejection temperature T_C is held at 30 °C.

Individual $NTU_{xe} = 2.0$, $NTU_r = 100.0$, $NTU_{xc} = 2.0$ are arbitrary – but representative of rms values back-calculated for bench-mark engines such as the GPU-3. Parameter values against individual curves are percentages of the notional maximum.

For Figure 8.2 temperature of heat reception is fixed at 600 °C. It is assumed that the gas path is potentially capable of some notional maximum thermal loading specified in terms of NTU (Number of Transfer Units – the index of steady-flow heat exchange capacity).

Plotting specific work $Power/(m'RT_C)$ against fractional dead-space ν reveals that reduced heat exchange capacity calls for increased dead-space (reduced compression ratio r_p) to compensate the adiabatic component of temperature rise and consequent overload of the exchangers. Where exchange capacity approaches 100% of notional maximum, decreasing ν (thereby increasing r_p) gives increasing specific output. This is consistent with expectation from the 'isothermal' analysis, which also assumes exchanger efficiency to be independent of thermal load.

For Figure 8.3 the expansion-space temperature is again held at 600 °C. Calculated output is this time plotted against kinematic displacement ratio λ. The parameter is again a percentage of nominal exchanger capacity.

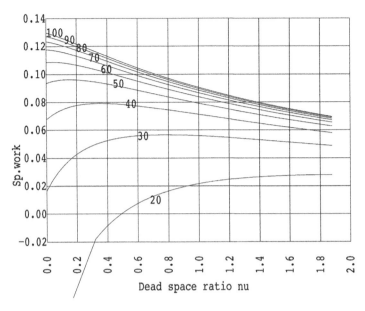

Figure 8.2 Specific cycle work as a function of dimensionless dead-space. The parameter is % notional maximum thermal load rating of exchangers

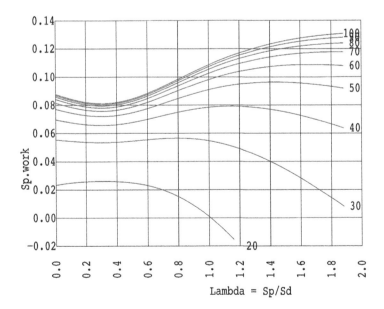

Figure 8.3 Specific work as a function of kinematic displacement ratio λ. The parameter is percentage of rated heat exchanger performance

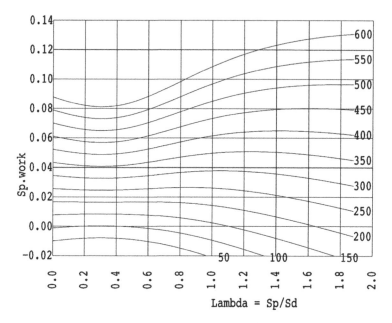

Figure 8.4 Specific work as a function of kinematic displacement ratio λ. Parameter is nominal temperature of heat reception T_E (°C)

On every curve there is a value of λ which maximizes specific performance. As heat exchanger performance reduces, so numerical values of 'optimum' λ fall below unity.

At first sight there is a remarkable 'coincidence' – near-duplication of Figure 8.2. Any mystery evaporates on recalling that the engine is driven by temperature *difference* $\Delta T/T_E = (T_E - T_C)/T_E$ [K] and that rate of convective heat exchange is the product $hA_w\Delta T$, where h is coefficient of convective heat transfer [W/m²K] and A_w [m²] is wetted area: a change of a given percentage in h (via change in *NTU*) is likely to have an effect similar to that of a change of comparable percentage in ΔT.

For Figure 8.3 exchanger performance remains set to 100% while heat reception temperature is T_E varied from 600 °C down to 50 °C.

Figure 8.4 suggests that getting the best out of a beta engine at decreasing T_E calls for decreasing kinematic displacement ratio λ. This is consistent with experimental findings by Larque and with the account by Karabulut et al. (2009). It is also in line with that of Lopez (2011), whose experiments suggested that varying swept volume ratio can change start-up temperature by more than 100 °C.

Figure 8.4 and Figure 8.1 (schematic representation of the cycle) recall the (open cycle) regenerative gas turbine where, in contrast to the non-regenerative case, efficiency peaks at low compression ratio.

8.5 Conclusions

Compression ratio r_v is a defining characteristic of reciprocating internal combustion engines – diesel and two- and four-stroke petrol engines. Chapter 2 found a numerical value of r_v to be essential in quantifying the sequence of ideal processes of the Carnot cycle.

Chapter 5 established a link between the r_v of a Stirling engine and its willingness to start, to generate power between 'intermediate' temperature differences, and for no-load running to persist after removal of the heat source.

This chapter makes use of appropriate (steady-flow) temperature solutions in the context of a closed regenerative cycle adapted for best possible representation of the defining features of the reciprocating Stirling engine. The calculated performance is again a strong function of r_v.

When the r_v of the steady-flow case are converted to their equivalent in terms of λ and β, the value pointing to best specific performance of the steady-flow engine is that which favours the reciprocating case.

8.6 Solution algorithm

Equations 8.1 to 8.8 are re-written:

$$T_2 = a_{11}T_1 \tag{8.9}$$

$$T_3 = a_{22}T_2 - b_2 \tag{8.10}$$

With a T_1 given, values for T_2 and T_3 follow.

$$T_4 = T_3 + T_7 - T_8 \tag{8.11}$$

$$T_5 = a_{44}T_4 - b_4 \tag{8.12}$$

$$T_6 = a_{55}T_5 \tag{8.13}$$

$$T_7 = a_{66}T_6 - b_6 \tag{8.14}$$

$$T_8 = -a_{73}/a_{78}T_3 - a_{74}/a_{78}\,T_4 - a_{77}/a_{78}T_7 \tag{8.15}$$

$$T_1 = a_{88}T_8 + b_8 \tag{8.16}$$

Numbering of the subscripted coefficients is a carry-over from a solution which called a FORTRAN library sub-routine for matrix 'inversion'.

Equation 8.15 is added to Equation 8.11, eliminating T_8:

$$(1 + a_{73}/a_{78})T_3 + (-1 + a_{74}/a_{78})T_4 + (1 + a_{77}/a_{78})T_7 = 0 \tag{8.17}$$

The coefficients of Equation 8.17 are abbreviated to c_3, c_4 and c_7. T_7 is then eliminated with the aid of Equation 8.14 to give T_4 in terms of T_3:

$$T_4 = \frac{c_7(b_6 + a_{55}a_{66}b_4) - c_3T_3}{c_4 + c_7a_{44}a_{55}a_{66}} \tag{8.18}$$

Values for T_5, T_6, T_7, T_8, and T_1 now follow in succession. If the starting value of T_1 was an estimate (for example, T_C), then the value according to Equation 8.16 will be somewhat different. The average of the two values is an improved T_1. The sequence iterates to convergence.

9

Dynamic Similarity

9.1 Dynamic similarity

Dynamic similarity is a powerful tool in design, testing and development of ship hulls, screw propellers, aircraft wings, vehicle bodies, and so on. It provides the accepted way of relating model behaviour in the wind-tunnel or towing-tank to performance of the full-size counterpart. The method is not confined to any particular environment – water, air, and so on. but has been late coming to the aid of Stirling engine design.

When indicated power Pwr [W] is generated by swept volume V_{sw} [m³], at rpm revolutions per minute and at mean charge pressure p_{ref}, the variables defining the gas process cycle are:

Geometry (drive mechanism kinematics and gas path dimensions):

$$\ell_1, \ell_2, \ell_3, \ldots \ell_n$$

Operating conditions:

$$p_{ref}, rpm, T_E, T_C$$

Working fluid properties:

$$R, c_p(\text{or } c_v \text{ or } \gamma), \mu_0, k_0$$

Subscript $_0$ implies a value at some agreed reference temperature, as for example, $\mu_0 \equiv \mu(T_0)$.

The additional variables ρ_w and α_w required to define the cycle history of transient temperature in the regenerator matrix have not been overlooked: they will be re-introduced later.

Power and usage of Buckingham's 'pi' theorem are well-known. Applying to the present case yields the following functional expression for dimensionless power (specific cycle work):

$$\frac{Pwr}{p_{ref}V_{sw}f} = N_B = f\{\ell_1/V_{sw}^{1/3} \ldots \ell_n/V_{sw}^{1/3},$$

$$T_E/T_C, \quad \gamma, \quad p_{ref}/\omega\mu, \quad \omega V_{sw}^{1/3}/\sqrt{(RT_C)}, \quad kT_C/(\omega p_{ref}V_{sw}^{2/3})\}$$

(9.1)

Stirling Cycle Engines: Inner Workings and Design, First Edition. Allan J Organ.
© 2014 John Wiley & Sons, Ltd. Published 2014 by John Wiley & Sons, Ltd.

Equation 9.1 makes use of the inter-changeability of *rpm*, cycle frequency f [s^{-1}] and angular speed $\omega = 2\pi rpm/60$ [rad/s]

The upper row of dimensionless lengths in Equation 9.1 represents a fundamental condition for Dynamic Similarity, namely that there should be geometric similarity: for two engines to have identical gas processes and identical specific cycle work, the one should be a scale model of the other – or a photocopy enlargement.

Alternative combinations of the groups in the lower row are possible. Let $p_{ref}/\omega\mu_0$ be abbreviated to N_{SG} and $\omega V_{sw}^{1/3}/\sqrt{(RT_C)}$ to N_{MA}. Then:

$$N_{SG}N_{MA}{}^2 = p_{ref}\omega V_{sw}{}^{2/3}/\mu_0 RT_C = N_{RE} \tag{9.2}$$

N_{RE} has the personality of a Reynolds number, and will be called Reynolds parameter.

Define group $kT_C/(\omega p_{ref}V_{sw}{}^{2/3})$ as N_K, and product $N_{RE}N_K$ is seen to be $N_{pr}{}^{-1}\gamma/(\gamma - 1)$, where N_{pr} is Prandtl number $\mu_0 c_p/k$. Numerical values of N_{pr} differ little between candidate gases, and γ is a parameter in its own right, so for present purposes the group is N_K redundant.

With N_T representing temperature ratio T_E/T_C, Equation 9.1 reduces to unexpectedly compact form given the context – cyclic, compressible flow of a gas with friction and intense unsteady convective heat transfer:

$$\frac{Pwr}{p_{ref}V_{sw}f} = f\{\ell_1/V_{sw}{}^{1/3} \ldots \ell_n/V_{sw}{}^{1/3}, N_T, \gamma, N_{SG}, N_{MA}\} \tag{9.3}$$

Equation 9.3 has potential for use in *scaling* – use of data from an engine of known performance (the *prototype*, 'prot') for generating a design or designs of different power [the *derivative(s)*, 'deriv'] of different charge pressure p_{ref}, of different working fluid, different *rpm*, different swept volume, and so on – but of un-changed gas process cycle, and thus of un-changed indicated thermal efficiency and indicated specific cycle work.

The general principle is obvious enough. Numerical values of individual parameters (γ, N_{SG}, N_{MA}, etc.) of the derivative are equated to those of the prototype, as for example, $N_B{}^{deriv} = N_B{}^{prot}$. Three instances are non-trivial:

$$N_B{}^{deriv} = N_B{}^{prot}$$
$$N_{MA}{}^{deriv} = N_{MA}{}^{prot}$$
$$N_{SG}{}^{deriv} = N_{SG}{}^{prot}$$

A numerical value is known for each right-hand element by calculation from prototype dimensions, operating conditions and measured performance. The designer will decide target power for the derivative and choose working fluid, thereby supplying numerical values of Pwr, R, c_p (or γ) and μ_0 required in evaluating the left-hand sides. The situation leaves three unknowns V_{sw}, p_{ref} and ω. Transferring to the right-hand sides symbols for quantities having numerical values pre-set in this way:

$$(p_{ref}V_{sw}\omega)^{deriv} = 2\pi Pwr^{deriv}/N_B{}^{prot} \tag{9.4}$$
$$(\omega V_{sw}{}^{1/3})^{deriv} = (RT_C)^{deriv}N_{MA}{}^{prot} \tag{9.5}$$
$$(p_{ref}/\omega)^{deriv} = \mu_0{}^{deriv}N_{SG}{}^{prot} \tag{9.6}$$

Equation 9.6 is substituted into Equation 9.4 to eliminate p_{ref}. Equation 9.5 is substituted into the result, yielding an explicit expression for V_{sw}^{deriv}. The resulting numerical value of the ratio $(V_{sw}^{deriv}/V_{sw}^{prot})^{\frac{1}{3}}$ is the linear scale factor S by which every linear dimension of the derivative is multiplied to yield an engine of target power Pwr^{deriv}:

$$S = 2\pi\{Pwr/(R\mu_0 \, T_C)\}^{deriv}\{V_{sw}^{\frac{1}{3}} \cdot N_B^{prot} \cdot N_{RE}^{prot}\}^{-1} \tag{9.7}$$

A value for ω^{deriv} – and thus for rpm^{deriv} – is now available from Equation 9.5:

$$\omega^{deriv} = (RT_C/V_{sw}^{\frac{1}{3}})^{deriv} N_{MA}^{prot} \tag{9.8}$$

With the numerical value for ω^{deriv} established, charge pressure p_{ref}^{deriv} follows from Equation 9.6 (or from Equation 9.4):

$$p_{ref}^{deriv} = (\omega\mu_0)^{deriv} N_{SG}^{prot} \tag{9.9}$$

9.2   Numerical example

General Motors' GPU-3 engine serves as prototype. Table 9.1 gives that part of the specification which is used in calculation. The numerical values for N_B, N_{MA}, and so on are based on the peak power point with H_2 as working fluid.

The attractions of a derivative using air or N_2 (rather the H_2 of the GPU-3) are self-evident. A specific power penalty is inevitable, and it might be realistic to expect 1 kW from a derivative of size envelope comparable to that of the GPU-3.

Design (or, more appropriately, re-design) is by evaluating Equations 9.7, 9.8, and 9.9 in that order. Each right-hand side can be thought of a function of two derivative variables: For Equation 9.7, $S = S\{Pwr, (R\mu)\}$, for Equation 9.8, $\omega = \omega\{R, V_{sw}\}$ and for Equation 9.9 $p_{ref} = p_{ref}\{\omega, \mu_0\}$. The reason for treating $R\mu_0$ as a single variable is that the choice of working gas has fixed the numerical value of both elements of the product.

On this basis each equation can be represented in a three-parallel-scale nomogram. The diagonal line on Figure 9.1 joins target power (1 kW) of the left-hand scale to the (built-in)

Table 9.1   Linear scale factor S is calculated from a small number of dimensionless parameters, listed here from published specifications of selected candidate prototypes

	N_T	γ	N_{MA}	N_{SG}	N_B
400 hp/cyl	3.08	1.66	0.0145	13.6E+09	0.235
GPU-3	3.27	1.4	0.0166	2.18E+09	0.1832
MP1002CA	2.92	1.4	0.0210	0.56E+09	0.1169
V-160	2.727	1.66	0.0121	5.62E+09	0.0948*
P-40	3.07	1.4	0.0191	4.27E+09	0.0840
PD-46	2.67	1.66	0.0168	1.90E+09	0.0764
Clapham 5 cc	3.07	1.66	0.0046	2.28E+09	0.0096

*Based on 8 kW output at rated *rpm* (1500).

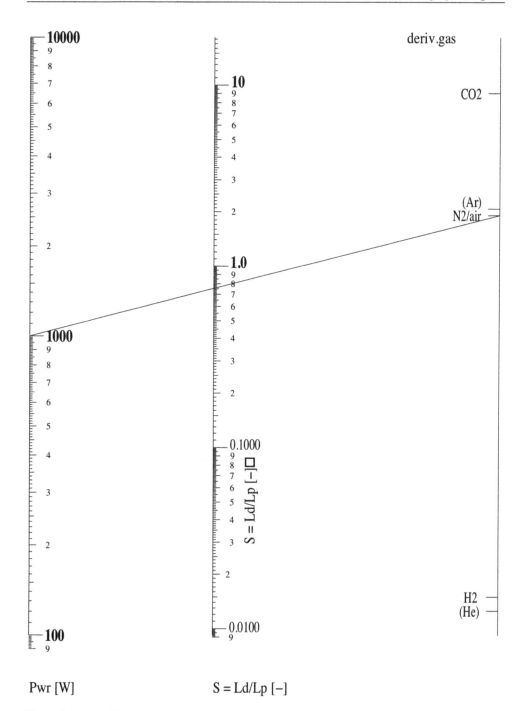

Pwr [W]   S = Ld/Lp [−]

Figure 9.1 Graphical equivalent of Equation 9.7 for the GPU-3 at peak power point on H_2. S is linear scale factor L_d/L_p

Vsw [cc]

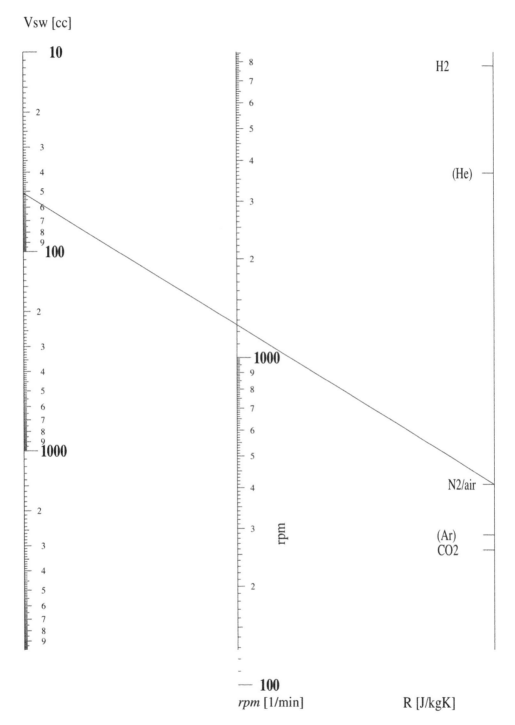

Figure 9.2 Graphical equivalent of Equation 9.8 for the GPU-3 at peak power point on H₂

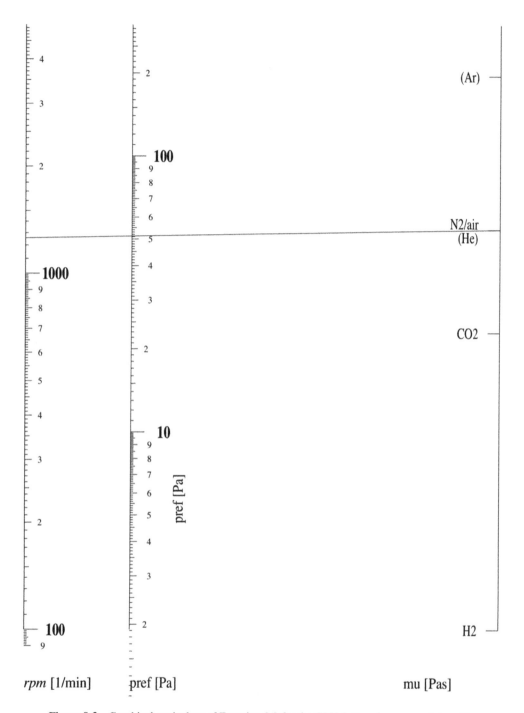

Figure 9.3 Graphical equivalent of Equation 9.9 for the GPU-3 at peak power point on H_2

Table 9.2   Gas path specification of 1 kW, air/N$_2$-charged machine scaled from the GPU-3 by Dynamic Similarity

p_{ref}	51.25	bar
rpm	1258.4	min^{-1}
V_{sw}	51.01	cm^3
L_{xe}	185.16	mm
d_{Txe}	2.28	mm
$(n_{Txe}$	40	$-$)
L_{xc}	34.8	mm
d_{Txc}	0.82	mm
$(n_{Txc}$	312	$-$)
L_r	17.06	mm
d_w	0.0302	mm
m_w	10.4	mm^{-1}

numerical value of $R\mu_0$ for air H$_2$ the chosen working fluid in this case. Linear scale factor S is read from the inner scale as 0.755 [$-$].

This sets swept volume ratio $V_{sw}^{deriv}/V_{sw}^{prot}$ to $0.755^3 = 0.430$. The swept volume V_{sw} of the GPU-3 is 118 cc, that of the derivative is $0.43 \times 118.0 = 50.74$ cc.

Solving Equation 9.8 can be carried out by entering the left-hand scale of Figure 9.2 at this value, *noting that graduations elapse downwards*. Numerical values of gas constant R are built in to the right-hand scale at the relevant label. Connecting left- and right-hand scale values as indicated gives derivative rpm as 1250 (1258.4 by hand calculator).

Use of Figure 9.3 to acquire derivative p_{ref} is now self-evident, the result being 51.25 bar. For $\gamma = 1.4$ design can go forward with these figures. Realization proceeds by multiplying every linear dimension of the GPU-3 gas path by scale factor S – exchanger tubes lengths L_x, tube internal diameters d_{Tx}, regenerator wire diameter d_w, and so on. (Mesh number m_w [wires/mm] is *divided*!) Exchanger tube numbers n_{Tx} are unaltered from prototype values.

For the GPU-3 expansion exchanger details are $L_{xe} = 245.3$ mm, $d_{Txe} = 3.02$ mm, $n_{Txe} = 40$. Table 9.2 gives the gas-path specification of the 1 kW, air-charged derivative.

9.3   Corroboration

If an engine design has already been derived by Dynamic Similarity principles, no account has come to the attention of the author. On the other hand, design methods are available which are independently formulated. One of these is *Scalit*, based on *energetic similarity* which dispenses with the requirement for geometric similarity. Energetic similarity adds flexibility by accepting power, *rpm* and swept volume of the proposed derivative as input. *Scalit* has access to gas path geometry and operating point of the GPU-3. Feeding values acquired via Figures 9.1, 9.2, and 9.3 leads to the gas path specification reproduced as Table 9.2.

For practical purposes the results of the two methods are indistinguishable. This might be considered remarkable considering that *Scalit* achieves energetic similarity by equating *NTU* (*Number of Transfer Units*) of the derivative to that of the prototype. Heat transfer and flow friction correlations ($StPr^{2/3} - Re$ and $Cf\text{-}Re$) embodied in the algebra are for steady flow,

Echo of input data – check carefully!!
****

Power/gas path: 1.000 KW
Reference pressure: 51.25 bar/atm
Rev/min: 1258 *rpm*

Scaled swept vol./gas cct: 51.05 cubic cm

Press <RETURN> or <INPUT> to continue output

	Expansion	Regenerator	Compression
Lengths mm	185.22	17.06	34.81
Free-flow area mm sq	0.163E+03	0.102E+04	0.162E+03
Hydraulic radii mm	0.570	0.023	0.204
Tube int. dia. mm	2.280		0.815
No. indiv. cyl. tubes	40	-	312
Regen. wire dia. mm		0.030	
Regen mesh 1/mm (1/in)		10.383 (263)	
Gauzes in stack		282	
Volumetric porosity		0.75	
X-sect area housing(s) mm.sq		0.136E+04.	

Figure 9.4   Screen image of *Scalit* session fed with with p_{ref}, *rpm* (and power) of the Dynamic Similarity study

as being the only such correlations available. Significant departures from the predictions of Dynamic Similarity might have been anticipated – but do not materialize.

9.4   Transient response of regenerator matrix

The additional variables required are density ρ_w, specific heat c_w and thermal conductivity k_w of the parent material of the matrix. Because ρ_w and c_w occur only in combination as $(\rho c)_w$, only two additional groups are anticipated: thermal capacity ratio $T_C(\rho c)_w / p_{ref}$, denoted N_{TCR} and Fourier modulus $N_F = \alpha_w / \omega d_w^2$, where α_w is thermal diffusivity, $k_w/(\rho c)_w$ [m^2/s].

For most practical purposes the numerical values of k_w, ρ_w and $c_w\rho_w$ – and thus of α_w – will be those for stainless steel. Although woven wire mesh is available in various grades of

Table 9.3   Parameters N_{TCR} and N_F characterize regenerator matrix transient response. The Table lists peak power point values

	N_{TCR}	N_F
400 hp/cyl	312	4.3
GPU-3	193	6.05
MP1002CA	876	15.92
V-160	88	10.19
P-40	87.7	3.86
PD-46	129	7.96
Clapham 5 cc	166	0.0398

rpm [1/min]

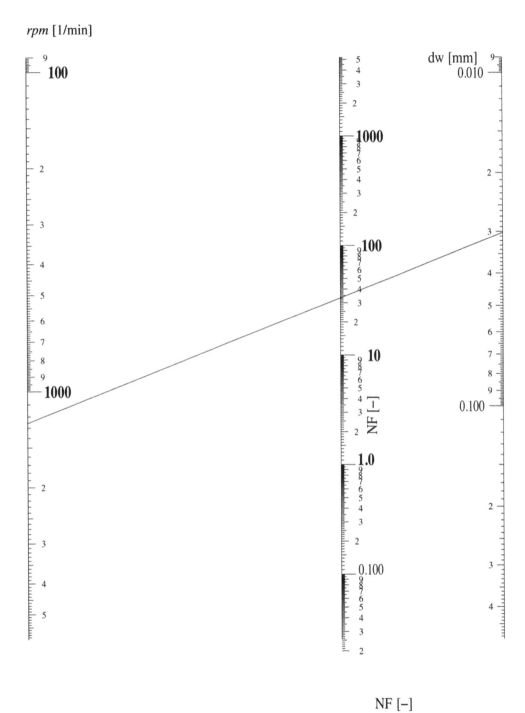

NF [−]

Figure 9.5　Graphical equivalent of definition of N_F. Constructed for matrix of stainless steel to AISI 301, and approximately valid for other AISI grades

stainless steel, the variation in thermal properties between grades is insignificant by comparison with variations in ω, p_{ref}, and so on. Adding either of the constraints $N_{\text{TCR}}^{\text{deriv}} = N_{\text{TCR}}^{\text{prot}}$ or $N_F^{\text{deriv}} = N_F^{\text{prot}}$ to the original three similarity conditions therefore results in the prototype at its operating point scaling to ... the prototype at its operating point!

The situation is dealt with pragmatically: the numerical value of N_{TCR} determines the maximum possible matrix temperature swing, assuming infinite local Nusselt number and unlimited α_w. The value of N_F determines the extent of penetration of the temperature wave over a cycle for finite α_w.

In the 'perfect' regenerator N_{TCR} and N_F are both infinite. An increase in the numerical value of either between prototype and derivative would be consistent with performance improvement. Few in number must be the designers who would be dissatisfied with departure from strict similarity if the result were performance enhancement. A scaling study therefore concludes by comparing N_{TCR} and N_F of prototype and derivative.

Table 9.3 is a continuation of Table 9.1 with entries for the N_{TCR} and N_F of candidate prototypes. Noting the definition of N_{TCR} it is necessary only to check that $p_{\text{ref}}^{\text{deriv}} < p_{\text{ref}}^{\text{prot}}$ to be sure that $N_{\text{TCR}}^{\text{deriv}} < N_{\text{TCR}}^{\text{prot}}$ – as is clearly the case here.

N_F^{deriv} requires calculation. Use of the nomogram of Figure 9.5 takes the effort out of the arithmetic, and yields $N_F = 33.3$ [–]. Being a factor of five larger than the prototype value the result suggests that regenerator transient response will be improved.

9.5   Second-order effects

Numerical values, such as that for coefficient of dynamic viscosity μ, k are temperature-dependent. To this point only, (fixed) reference values μ_0, and so on. have been used, whereas gas temperature T throughout the gas path varies through a factor of about three.

This is of no consequence provided temperature-dependence expressed as $\mu(T)/\mu(T_0)$ is similar for prototype and derivative gases. Between He and N_2/air this is the case.

Ironically, strict similarity is unachievable between the former (monatomic) and latter (diatomic) gases. Coincidence between the scaleable gases H_2 and air/N_2 is less good.

9.6   Application to reality

A specification generated by scaling will not necessarily make for a viable project: operational characteristics (p_{ref}, rpm) may be unacceptable, and/or fabrication of scaled number n_{Tx} of tubes may be too challenging. (Try scaling the GPU-3 to the power level of Philips' 400 hp/cylinder concept!)

Even if the scaled specification is attractive it cannot be an optimum unless that of the GPU-3 an optimum. Conversely, if the resulting engine fails to perform, the fault will almost certainly lie with engineering implementation rather than with the specification per se.

Where the derivative specification appears promising, this is not the end of the design study: External heat supply and removal must be such that the internal temperatures which prevail in the prototype are achieved in the derivative.

10

Intrinsic Similarity

10.1 Scaling and similarity

The previous chapter showed that *rpm* and charge pressure p_{ref} for an engine scaled up or down geometrically may be so chosen that the cycle of gas processes remains unaltered. This gives every reason to expect indicated efficiency and (indicated) Beale number N_B to remain as for the prototype. However, power will be that which results when N_B is multiplied by p_{ref} and *rpm*/60.

More commonly, it is a specific value of power which is the design target. The figure for swept volume V_{sw} is arrived at by exploring candidate *rpm* and/or charge pressure. The process can be formalized in at least three variants of scaling. The latter process is essentially a routine for achieving, in a 'derivative' design, the cycle gas process history of a 'prototype', overcoming differences in size and/or working fluid and/or *rpm* in the process. Chapters 11, 12, and 13 each focus on one such process.

Intimate details of the cycle of gas interactions have never been witnessed – and probably never will be. An attraction of scaling is that it dispenses with the need for any such insight. On the other hand, the arithmetic of scaling applied with a degree of inside information is less likely to go off the rails than the identical process applied by rote. The previous chapter introduced *Similarity principles* via the traditional route. All variants of scaling exploit Similarity to one level of rigour or another, so what may shed light is an alternative, intuitive – rather than algebraic – perspective: a Stirling engine is, after all, a Stirling engine, converting heat to work just as did the 1818 prototype – and every derivative since. To that extent, similarity (with lower-case 's') is intrinsic.

10.2 Scope

To embrace *all* Stirling engines would be a tall order, but manageable coverage is achieved if attention is restricted to a 'sub-set' of the *genre*.

A promising 'sub-set' remains on discarding free-piston types, table-top demonstrators and 'low delta-T' engines. Those which remain have kinematic volume phasing (Walker's 'disciplined pistons' 1980), multi-channel exchangers and micro-porous regenerator. Operation

Stirling Cycle Engines: Inner Workings and Design, First Edition. Allan J Organ.
© 2014 John Wiley & Sons, Ltd. Published 2014 by John Wiley & Sons, Ltd.

is between temperatures T_E and T_C in the region of 650 °C and 35 °C respectively, giving characteristic temperature ratio N_T of about 3. Internal pressurization is offset by crank case or buffer space pressure.

The resulting degree of formal 'similarity' survives differences in configuration (coaxial 'beta' vs opposed-piston 'alpha') – and, obviously, number of cylinders. Similarity in this sense is clearly not visual. Rather, it lies in the close numerical agreement observed ...

> when data describing two or more machines (geometry, operating conditions) are substituted into the set of (dimensionless) parameters describing the gas processes resulting in respective values which are numerically close.

The power of this numerical tool can be mobilized directly without the need for derivation. On the other hand, a parameter system which is inherently 'right' overlaps other approaches to the same problem, as the following section attempts to show.

10.2.1   Independent variables

As far as analysis and modelling are concerned, gas process events are a function of time t [s] and location x [m]. That there is only one length coordinate x (rather than the x, y, z which might be expected) reflects the fact that data available for heat transfer and flow design (the 'steady flow correlations') assume one-dimensional or 'slab' flow. In this way, an expansion exchanger tube folded into a hairpin of leg length L has the same effective flow passage length L_{xe} as a straight tube of length $2L$.

L_{xe} for the Philips 400 hp/cylinder concept was 1090 mm. For Clapham's 5 cc engine it was 127 mm. No sign of similarity so far! In neither case is the length arbitrary: it has something to do with the 'thermodynamic size' of the engine, and the latter is characterized by *swept volume* V_{sw}. This suggests using 'reference length' $L_{ref} = V_{sw}^{1/3}$ [m] to pro-rate or 'normalize' absolute lengths for purposes of numerical calculation. Values of $L_{xe}/L_{ref} = \lambda_{xe}$ for the 400 hp/cylinder concept and the Clapham engine are respectively 4.39 [−] and 3.947 [−]. The difference is now 10% reckoned using the higher value.

The location of a gas process event is now defined in terms not of x [m] but of $\lambda = x/L_{ref}$ [−].

An 'event' such as mass rate dm/dt at location x and time instant t is proportional to *rpm* (zero *rpm*: zero mass rate) and thus a function of x, t, and *rpm*, viz, $dm/dt = f(x, t, rpm)$. It is therefore also proportional to angular speed ω, where $\omega = 2\pi rpm/60$ [rad/sec]. Dividing dm/dt by ω cancels the proportionality, thereby reducing the dependency by one variable. The new independent variable ωt is crank angle φ'. Taking account of the preceding paragraph, $dm/d\omega t = f(\varphi, \lambda)$.

Future formulations will be in terms of crank angle φ^1 and dimensionless length λ.

---

[1] Where compressibility effects dominate, graphical display of solutions acquired by use of the Method of Characteristics (MoC) is more readable if t is made dimensionless by dividing by time for a wave traverse at reference acoustic speed, that is, by plotting in terms of $t/(L/c_0)$.

10.2.2   Dependent variables

The mass of gas m instantaneously occupying any separately-identifiable volume or sub-volume V is given in terms of the gas equation $pV = mRT$, where R [J/kgK] is the gas constant for gas in question. The cyclic pressure swing in an un-pressurized engine might be between 0.6 bar and 1.4 bar. That of an engine pressurized with H2 might be between 60 bar and 140 bar. Expressed as a fraction of charge pressure p_{ref}, the cyclic swing in *specific pressure* $\psi = p/p_{ref}$ is between identical values – 0.6 [–] and 1.4 [–] in both cases.

More than three orders of magnitude separate the dead space V_{dr} of the regenerators of Philips' 400 hp/cylinder concept from that of the prize-winning Clapham 5 cc engine. As a fraction of swept volume V_{sw} (i.e., expressed as *dead volume ratio* $\delta_r = V_{dr}/V_{sw}$) the difference is a mere 15%.

Nominal heat rejection temperature T_C is almost always ambient. When a temperature at given location in the gas path and at a given point in the cycle is expressed as a fraction of T_C, viz., as $\tau = T/T_C$ variations between numerical values for different engines are minimized.

The ideal gas equation can be re-written:

$$p/p_{ref} \cdot V/V_{sw} = \frac{mRT_C}{p_{ref}V_{sw}} T/T_C$$

In terms of the new dependent variables:

$$\psi\delta = \sigma\tau$$

The term σ is *specific mass*:

$$\sigma = mRT_C/p_{ref}V_{sw} \tag{10.1}$$

By the earlier reasoning, specific mass *rate* σ' is:

$$\sigma' = mRT_C/\omega p_{ref}V_{sw} \tag{10.2}$$

Figure 10.1a is constructed on the basis of the specification of Philips' 400 hp/cylinder design – kinematics, geometry, working fluid, and operating conditions, together with the assumptions of the 'isothermal' analysis. Specific mass rate σ' is plotted at four strategic locations in the gas path. The picture is idealized – but to exactly the same extent as that for two other engines – the GPU-3 (Figure 10.1b) and the V-160 (Figure 10.1c).

Calculating in terms of σ' has evidently gone a long way to absorbing differences in mechanical configuration, working fluid and operating conditions – *rpm* and charge pressure p_{ref}. The size range spanned, in terms of swept volume V_{sw}, is in the impressive ratio of 130:1! With diagram (*a*) available, there is barely a need to plot (*b*) and (*c*) – and no need to re-plot for any other Stirling engine falling within the category.

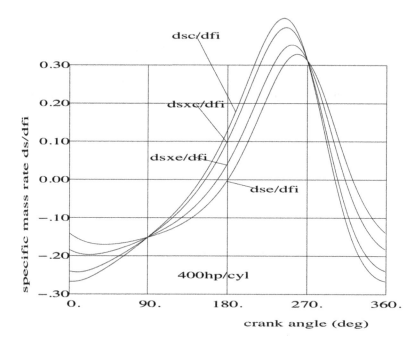

Figure 10.1a   Variation of specific mass rate σ′ with crank angle φ for Philips 400 hp per cylinder design at four locations: e, expansion cylinder entry/exit; xe, hot end of regenerator; xc, cold end of regenerator; c, compression cylinder entry/exit

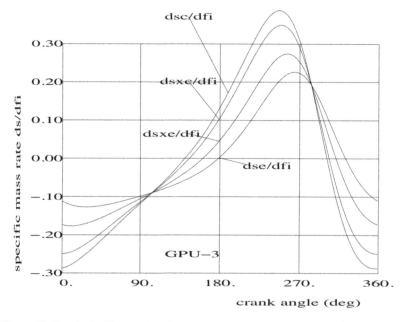

Figure 10.1b   As for Figure 10.1a, but constructed for General Motors' GPU-3 engine

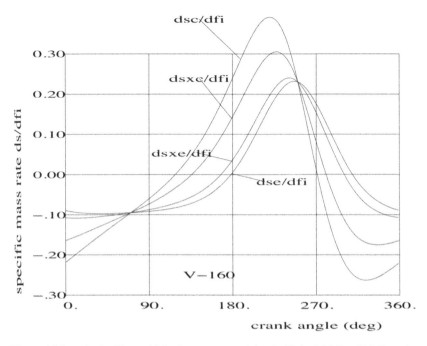

Figure 10.1c   As for Figure 10.1a, but constructed for the United Stirling V-160 engine

10.2.3   Local, instantaneous Reynolds number Re

For flow at speed u [m/s] in a duct of hydraulic radius r_h [m] at temperature close to ambient where coefficient of dynamic viscosity is μ_0 [Pas][2] Reynolds number Re is:

$$Re = \frac{4\rho\,|u|\,r_h}{\mu_0}$$

Denoting local free-flow area by A_{ff} [m^2], noting that $\rho|u| = |m'|/A_{ff}$ and substituting for σ' from Equation 10.2 allows re-writing as:

$$Re = 4r_h\;\frac{\omega p_{ref} V_{sw}}{A_{ff} R T_C \mu_0}\,|\sigma'|$$

Breaking down V_{sw} into the product $V_{sw}^{2/3} V_{sw}^{1/3}$, making the substitutions and multiplying numerator and denominator by ω:

$$Re = 4\frac{r_h V_{sw}^{1/3}}{A_{ff}}\;\frac{p_{ref}}{\omega \mu_0}\;\frac{\omega^2 V_{sw}^{2/3}}{R T_C}\,|\sigma'| \qquad (10.3)$$

---

[2]Temperature dependence of μ is accounted for by using a reference value μ_0 multiplied by a dimensionless function $f(T)$ of temperature, viz., $\mu = \mu_0 f(T)$.

$V_{sw}^{2/3}$ is equal to L_{ref}^2, so that the second and third groups of Equation 10.3 are respectively Stirling parameter N_{SG} and the square of speed parameter N_{MA} introduced in Chapter 9.

$$Re = 4|\sigma'| \frac{r_h V_{sw}^{1/3}}{A_{ff}} N_{SG} N_{MA}^2 \qquad (10.3a)$$

Numerical values for A_{ff} and r_h can be acquired for an exchanger of any cross-section. The exchanger comprising a bundle of n_{Tx} parallel cylindrical tubes each of internal diameter d_{Tx} [m] is a special case:

$$Re = 4|\sigma'| N_{SG} N_{MA}^2 \frac{1}{\pi n_{Tx} d_{Tx}/L_{ref}} \qquad (10.4)$$

Equation 10.4 has made use of $r_h = \frac{1}{4}d_{Tx}$ for the cylindrical tube.

The more general case can be dealt with by noting that, whatever the cross-sectional geometry, the product $L_x A_{ff}$, is exchanger dead volume V_{dx} and $L_x A_{ff}/V_{sw}$ the dimensionless counterpart δ_x:

$$Re = (4/\delta_x)(L_x/L_{ref})^2 (r_h/L_x) N_{RE} |\sigma'| \qquad (10.5)$$

10.3   First steps

Whether design is to proceed by 'cut and try' experiment, or with the aid of computer simulation, or by scaling or – indeed – by intuition, it is normal to set a target shaft power Pwr. Charge pressure p_{ref}, swept volume V_{sw} and rpm will be a mutual trade-off. Choices may have to be re-visited when working fluid is settled on.

The Beale number criterion acts as 'practicality screening check' for the preliminary $V_{sw} - p_{ref} - rpm$ balancing act. It does so by excluding certain combinations[3] which violate practical experience. The check is carried out by applying the hand calculator to the inequality expression at Equation 10.6:

$$\frac{60.0 \times Pwr\,[\mathrm{W}]}{p_{ref}\,[\mathrm{Pa}] \times V_{sw}\,[\mathrm{m}^3] \times rpm\,[\mathrm{min}^{-1}]} \leq 0.15 \qquad (10.6)$$

There are at least three advantages to applying the check with the aid of the nomogram at Figure 10.2:

1. Doubts as to appropriate units (bar vs, cm^3 vs litre, kW vs W, etc.) do not arise.
2. A chart of this sort amounts to instant inversion of the underlying formula: Equation 10.6 has to be transposed to make the choice of parameter explicit. In the chart any choice is explicit.
3. Speed of calculation: after initial familiarization, a new combination of values takes seconds to verify.

---

[3]But not all! Pitfalls remain: the combination of wire mesh regenerator with air or N_2, for example, can lead to crippling compressibility effects. (Texts by this author appear to be alone in cautioning against and explaining – see Chapter 17.)

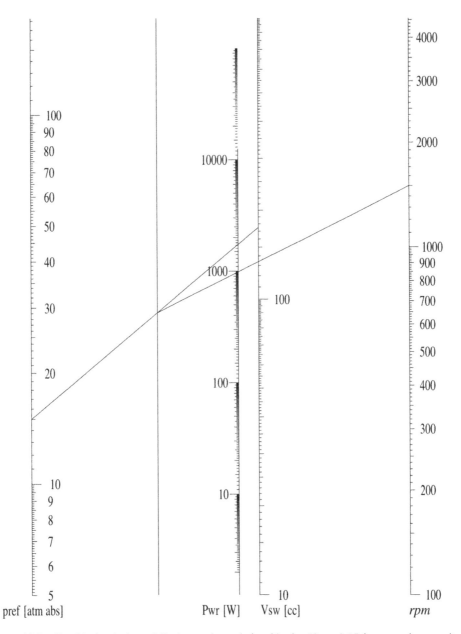

Figure 10.2 Graphical solution of Beale number relationship for $N_B = 0.15$ in convenience units. Connect p_{ref} scale [atm] *only* with V_{sw} scale [cm³] or vice versa and mark intersection with un-graduated scale. From this mark draw a straight line *either* to Pwr scale [W], noting intersection with *rpm* scale, *or* to *rpm* scale noting intersection with Pwr scale

Using the chart:

1. Join with a straight line *EITHER* p_{ref} to V_{sw} (or vice versa) *OR rpm* to *power* (or vice versa). Extending the straight line if necessary, mark the intersection with the auxiliary (ungraduated) vertical line.
2. From a point on the auxiliary scale draw straight line to the required value on *either* of the pairs of scales *not* selected for step 1. Intersections at the four graduated scales are numerical values of p_{ref}, V_{sw}, *rpm*, and power (in *convenience* units) which satisfy the Beale number N_B which has the (conservative) numerical value 0.15 when evaluated in consistent units.

If you have just used a four-scale alignment nomogram for the first time, check the result by using the formula.

Converting the resulting tentative values of V_{sw}, p_{ref}, and *rpm* to dimensionless operating parameters will allow gas path design to be pursued in detail *whilst temporarily postponing choice of working fluid!* Again, the calculator can be applied to respective algebraic expressions, viz, to $N_{MA} = \omega V_{sw}^{1/3}/\sqrt{RT_C}$, Stirling parameter $N_{SG} = p_{ref}/\omega\mu_{ref}$. Figures 10.3, 10.4 and 10.5 are respective nomogram equivalents. Scales are again calibrated in convenience units.

10.4   ... without the computer

In Equation 10.5 the product $N_{SG}N_{MA}^2$ has been abbreviated to the Reynolds parameter N_{RE}:

$$N_{RE} = N_{SG}N_{MA}^2 \tag{10.7}$$

Figure 10.6 is the equivalent nomogram. With ρ substituted for p/RT_C Equation 10.7 re-expands to

$$N_{RE} = \frac{\rho(\omega L_x)L_x}{\mu_0} \tag{10.8}$$

With ωL_x as characteristic velocity N_{RE} has the personality of a Reynolds number.

For given operating conditions and gas path location the only variable in Equations 10.4 and 10.5 is σ'. On the evidence of Figure 10.1, the variation of σ' with crank angle φ at any given gas path location is predictable. Interesting possibilities arise.

For the turbulent flow conditions assumed (Chapter 6, Section 6.1), the Stanton number *St* relates to *Re* as $StPr^{\ell} = aRe^{-b}$. The Prandtl number *Pr* being invariant for practical purposes, local instantaneous *NTU* value follows as $NTU = StL_x/r_h$. Substituting into Equation 10.5:

$$NTU \approx a\{4/\delta_x\,(L_x/L_{ref})^2\,N_{RE}\}^{-b}\,(L_x/r_h)^{1+b}\,|\sigma'|^{-b} \tag{10.9}$$

The availability of a value for *NTU* permits calculation of local, instantaneous heat transfer intensity – and, via the concept of lost available work, corresponding penalty due to imperfect

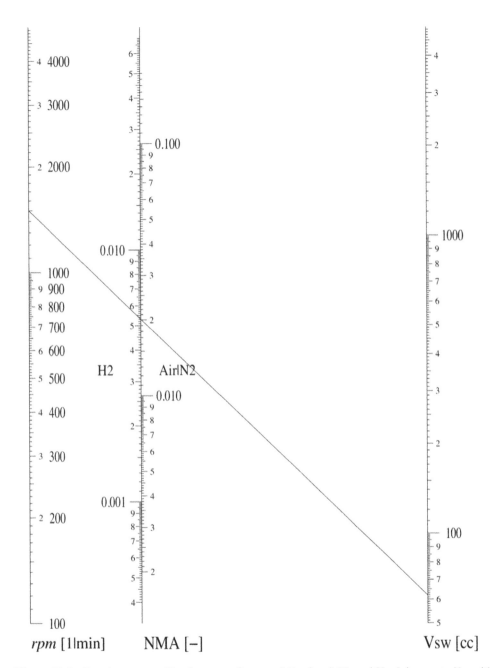

Figure 10.3 Speed parameter N_{MA} in terms of *rpm* and V_{sw} for air/N_2 and H_2. Join *rpm* to V_{sw} with straight line and read off N_{MA} from centre scale

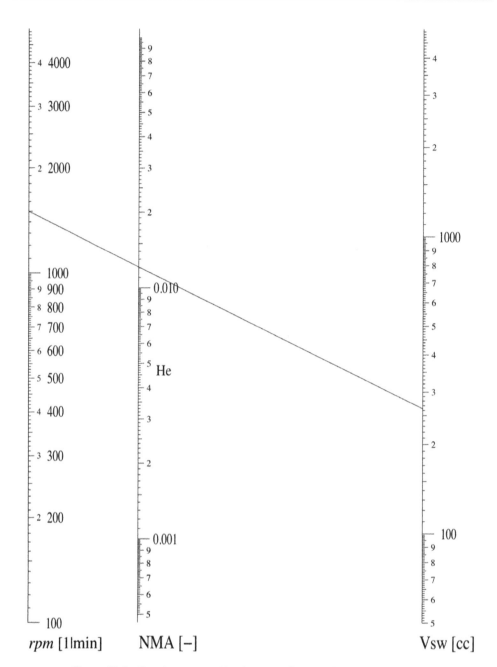

Figure 10.4 Speed parameter N_{MA} in terms of *rpm* and V_{sw} for helium (He)

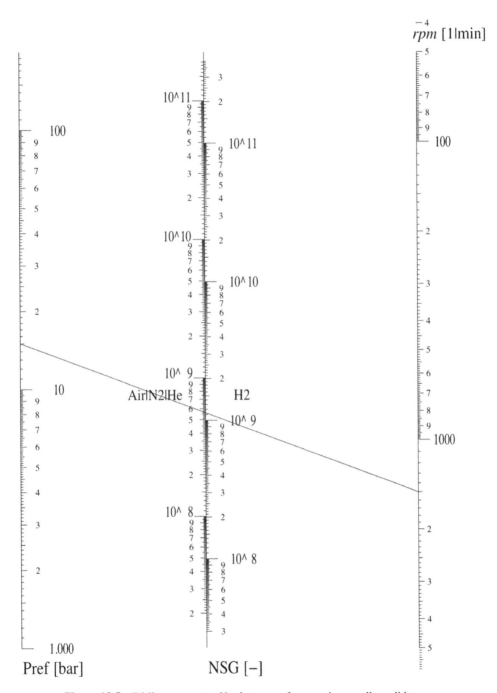

Figure 10.5 Stirling parameter N_{SG} in terms of rpm and p_{ref} – all candidate gases

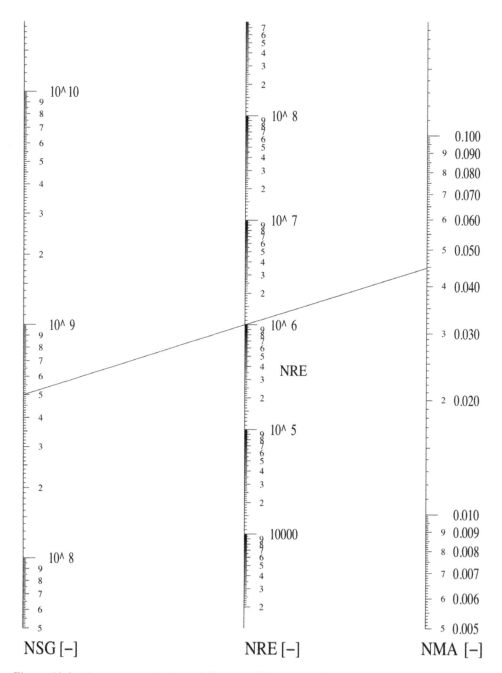

Figure 10.6   Nomogram equivalent of Equation 10.7: Reynolds parameter N_{RE} in terms of N_{SG} and N_{MA}

heat transfer. Integration with respect to $d\varphi$ of $|\sigma'|^{-b}$ between limits of 0 and 2π leads to a constant for any chosen gas path location, and thus for the expansion exchanger, for example. The constant amounts to a *cycle invariant* – to a *CI*.

Expressing lost work as a fraction of the cycle ideal allows the designer to opt for, say, a 5% penalty at a chosen operating point (value of N_{RE}). The exchanger design can now be any combination of respective numerical values of δ_x, L_x/L_{ref}, and L_x/r_h which balance the equation. The nomogram is the ideal medium for exploring the various options, since candidate combinations are read simultaneously against a straight-edge moved over the scales.

The procedure has been formulated in no fewer than four independent ways, with independent coding for each – and always with the same disconcerting result: when applied retrospectively to the exchanger designs of benchmark engines (GPU-3, V-160, etc.), losses due to imperfect heat transfer exceed those due to flow friction (pumping) by two orders of magnitude. While the two types of loss are not routinely separated experimentally, the result can hardly represent reality.

The outcome of the integration leading to given CI is dependent on the numerical value of exponent b. The net value of the right-hand side of Equation 10.9 evidently depends on both a and b. If the a and b of the steady-flow correlation $StPr^c = aRe^{-b}$ mis-represent conditions in the gas path element over a cycle then an anomalous outcome is to be expected.

The computer implementations accept numerical values of a and b as input data. Logical and algebraic skeletons are unaffected by changes to data values. Progress on this otherwise promising front awaits re-acquisition of heat transfer and flow friction data under representative oscillating flow conditions.

11

Getting started

11.1 Configuration

Only the most basic design problems lend themselves to explicit or 'serial' solution. The rest are to varying degrees 'chicken and egg', requiring one or more iterations to minimize the compromise involved in reconciling the conflicting priorities. To this extent it matters little whether the start point here is perusal of geometric and kinematic options, or exploration of thermodynamic potential by abstract analysis.

A matter on which the designer has almost certainly decided from the outset is that of configuration – opposed-piston (alpha) versus parallel-coaxial (beta) versus gamma. The choice will to some extent determine which drive mechanisms are candidates. This in turn raises the matter of thermodynamic[1] volume ratio κ and thermodynamic[1] phase angle α. Strictly speaking, a numerical value of α has meaning only in the context of simple-harmonic motion (SHM) – which few drive mechanisms other than the Scotch yoke deliver. For most viable mechanisms it will be possible to identify a maximum and a minimum of α over a complete revolution, and thus a 'mean' for the cycle. The target mean value is important, and extant 'optimization' charts are no guide to the appropriate choice.

The following is an attempt to reconcile the findings of recent bench tests with those of gas process modelling and theoretical deliberations:

Single–gas–path engine charged to a few atm. with air/N$_2$

> *Target 1*: sophisticated heat exchangers and regenerator; low-friction seals; rolling element bearings; high temperature ratio ($N_T > 3$). The specification can probably exploit high compression ratio: $\kappa = 0.8$, $\alpha = 90$ degrees (Figure 11.1). Equivalent λ and β are 1.275 and 38 degrees respectively.

---

[1] In the opposed-piston machine driven in simple harmonic motion (SHM), kinematic volume ratio κ is identical to thermodynamic volume ratio. Kinematic and thermodynamic phase angles α are likewise identical. The kinematic volume ratio λ and kinematic phase angle β of gamma machines driven in SHM have *equivalent* κ and α which can be read from Figure 4.2, or calculated from Equations 4.2–4.5.

Stirling Cycle Engines: Inner Workings and Design, First Edition. Allan J Organ.
© 2014 John Wiley & Sons, Ltd. Published 2014 by John Wiley & Sons, Ltd.

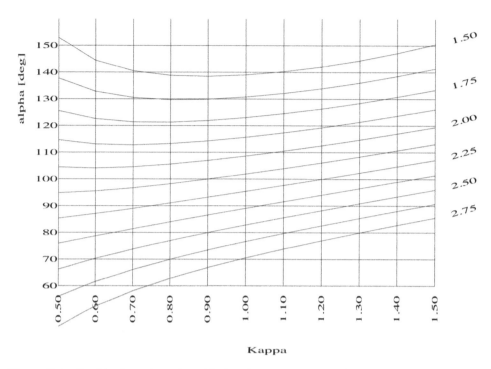

Figure 11.1   Combinations of κ and α achieving given compression ratio r_v (parameter curves). Constructed for dimensionless dead space $\nu = 1.5$

Target 2: basic heat exchangers and regenerator. Plain bearings. $N_T < 3$; small dead-volume. Choosing low compression ratio will minimize mechanical friction and need for the exchangers to deal with the thermal load resulting from increased adiabatic heating. Set $\kappa = 0.8$ again, but $\alpha = 130$ degrees Equivalent λ and β are 0.775 and 52 degrees respectively.

A multi-cylinder version embodying this option could be in 'Rinia' configuration, but with three inter-connected cylinders rather than four, giving $\alpha = 120$ degrees Any desired κ less than unity is achieved in terms of the diameter of the drive rod penetrating the compression space.

Engine charged to high pressure with hydrogen or helium.

The detail of the design is probably best generated by scaling from an existing engine (the 'prototype'). If drive kinematics are scaled linearly the derivative will have the volume ratio and phase angle of the prototype.

11.2   Slots versus tubes

From the point of view of heat transfer capability *internal* to the gas path there is little to choose between slots and tubes. The universal choice of tubes for engines of high power density (i.e., those charged to high pressure with hydrogen or helium) is the multi-tube arrangement, reflecting the difficulty in matching *NTU* arising internally to values achievable between combustion gases and the *external* surfaces: any mismatch amounts to a thermal bottleneck.

The potential mismatch is more tractable in the case of air- or N_2-based designs charged to modest pressure. This permits a choice between fins and tubes which will probably be decided on the basis of manufacturing convenience. Neither is an obvious candidate for mass manufacture, but there is an approach to the production of multiple slots which might change prospects.

It follows from observation of production methods for the races of rolling element bearings, some of which are now manufactured by wrapping from strip and butt-welding before finish-machining. Figure 11.2 suggests a sequence for adapting to the generation of internal slots.

The machining phase anticipates the final stage which is to extrude the wrapped and welded ring through a tapered die to achieve final external diameter simultaneously with

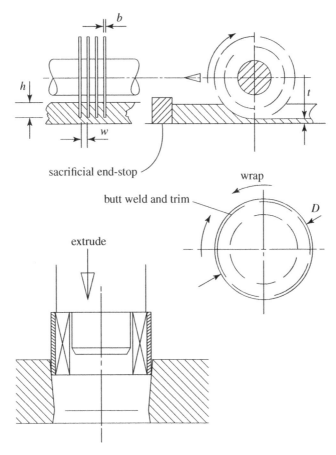

Figure 11.2   Schematic of sequence for production of internally-slotted exchanger achieving slot width *b* narrower than that of the tool which originally cut it

predetermined reduction in slot width w. The over-length strip has width equal to eventual slot length and thickness equal to target fin height h plus target pressure-wall thickness t *minus* a percentage for the increase which will occur during extrusion (expression below). If, as is likely, the aim is the largest number of slots of minimum practicable width, the strip is milled using a gang of slit-saws of the minimum thickness which will cope with the stock material – probably one of the stainless-steel grades.

Deformation due to the wrapping phase will cause slight distortion of the slots from parallel to cropped-triangular (trapezoidal). Regardless of final shape, the slot will have a calculable value of hydraulic radius for use in evaluation of Reynolds number Re.

Further deformation occurring during subsequent extrusion may be expected to be confined to circumferential compression *between* the fin roots, and not to include the fins themselves. This brings about further reduction in effective b and in increase t. A little algebra predicts the percentage reduction in diameter D required to achieve a given target reduction in b.

With reference to the notation of Figure 11.2:

$$n_s(w + b) = \pi D$$

$$b = \pi D / n_s - w$$

Differentiating:

$$db = \pi dD$$

$$db/b = \frac{\pi}{n_s(\pi D/n_s - w)} dD$$

$$db/b \, [\%] = dD/D \, [\%] \frac{1 + b/w}{b/w} \qquad (11.1)$$

Figure 11.3 expresses Equation 11.1 in the form of a design chart.

Cross-sectional area bt [m^2] remains constant during extrusion, allowing the change in t to be expressed in terms of that in b. Differentiating:

$$bdt/dD + tdb/dD = 0, \text{ or}$$

$$dt/t \, [\%] = -db/b \, [\%] \qquad (11.2)$$

Some hundreds of tubes per gas path are needed to reconcile wetted-area and hydraulic radius requirements at the compression end. At the assembly stage the sheer number is a problem in itself. Over and above this, failure of a single tube to braze or solder and to form a leak-free assembly renders the exchanger useless.

This problem is not inherent to the slotted exchanger. Moreover, a single slot can substitute a number of tubes.

The design stage would be expedited by a means of comparing multi-tube and slotted designs in terms of thermal performance *internal* to the gas path.

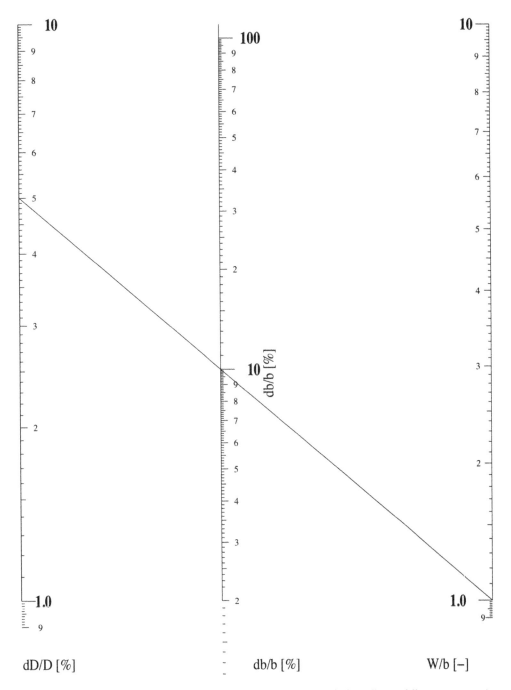

dD/D [%]   db/b [%]   W/b [−]

Figure 11.3 Equation 11.1 in design chart form. The specimen solution (diagonal line) expresses the self-evident fact that for slots and fins originally symmetrical ($b = w$), a 5% reduction in D brings about at 10% reduction in peripheral length element b which accommodates 100% of the plastic deformation per fin

11.3   The 'equivalent' slot

The comparison pre-supposes that the conduction path of the slotted exchanger and the multi-tube arrangement offers similar thermal resistance between gas and external coolant. (A specimen check appears later in this chapter.) The conversion algebra envisages number n of tubes, each of internal diameter d, being replaced by a single slot of width b and height h, as in Figure 11.4.

A first condition for similarity of thermodynamic performance is equivalence of dead space: If the slot and each of the n tubes it replaces are of the same length L_x in the axial (flow) direction:

$$\tfrac{1}{4}n\pi d^2 = bh$$

In the flow correlations of Kays and London (1964) the length variable in the definition of Reynolds number Re is *hydraulic radius* r_h. The choice goes some way to absorbing differences between flow correlations ($C_f - Re$ and $StPr^{2/3} - Re$) for cylindrical and rectangular slot. This justifies setting r_h of the slot ($\approx \tfrac{1}{2}b$ for $h \gg b$) equal to that of the tube, viz to $\tfrac{1}{4}d$:

$$b = \tfrac{1}{2}d \tag{11.3}$$

Substituting into the previous equation:

$$h = \tfrac{1}{2}\pi n d \tag{11.4}$$

Figure 11.4 is the graphical equivalent of Equation 11.3, and is in convenience units of mm. (Only integer values are of significance on the right-hand vertical scale for number of slots n.)

The inclined straight line represents a specimen conversion calculation. Tube internal diameter is 1.0 mm, and it is proposed to absorb tubes into a single slot. Connecting $d = 1.0$ of the left-hand scale to $n = 5$ of the right-hand scale gives $h = 7.8$ mm at the centre scale. From Equation 11.3, b for the slot is 0.5 mm.

If the multi-tube exchanger requires 250 tubes, the equivalent number of slots is $250/5 = 50$.

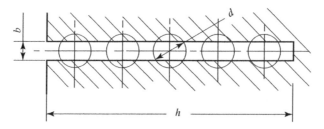

Figure 11.4   Approximate condition for equivalence – notation

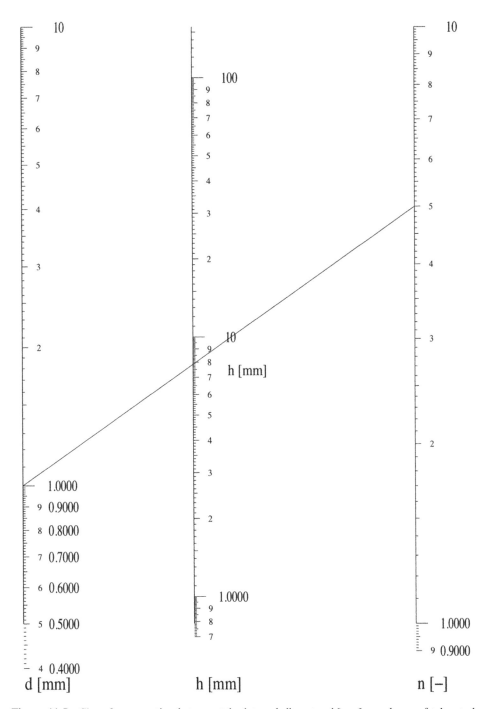

Figure 11.5 Chart for converting between tube internal diameter d [mm], number n of tubes to be replaced by a single slot and height h of the single slot

11.4   Thermal bottleneck

Modelling the gas processes can be approached in several ways. One option involves the use of '*perturbation*' – an imposing-sounding word for a simple concept. For present purposes, it means starting with a known approximation (the Schmidt analysis is a candidate) and calculating in terms of the (initially unknown) differences, or '*error*' terms – the ε – between the approximation and a more sophisticated statement of the gas exchange processes. The eventual solutions for the ε terms are added algebraically (i.e., with account of sign, $+/-$) to the starting approximation to give a more accurate picture.

There is a chain-reaction of advantages: for the temperature calculation, the error term $\varepsilon_T = T - T_w$ is the driving force of the local convective heat exchange process. A known value of ε_T is quickly converted via the concept of Availability Theory to an explicit value of *lost available work*.

And the chain reaction continues: Availability algebra reduces all losses (hydrodynamic pumping, defective heat exchange) *to a common denominator*, allowing them to be summed to give net performance penalty. The simple algebra covers thermal conduction through heat exchanger walls (where, however, the temperature difference driving heat flow through the solid is more appropriately denoted ΔT).

There is no catch! Numerical values of ε_T (and ΔT) resulting from application of Availability algebra may, however, come as a surprise. It will be easier to 'sell' this elegant, labour-saving approach after discussing a simple case showing that it yields a numerical result identical to that returned by traditional arithmetic.

Availability Theory elegantly quantifies *lost available work*: The upper-left diagram of Figure 11.6 indicates flow of heat by conduction at rate q' between a source at T_E and a sink at T_C. Entropy rate as q' [W] enters at T_E is (by definition) q'/T_E [W/K]. As q' leaves, entropy rate has increased to q'/T_C W/K – increased because, although q' is unchanged, $T_C < T_E$. A process of thermal conduction has generated entropy at rate s':

$$s' = q'\{1/T_C - 1/T_E\}W/K.$$

Allowing heat to be downgraded by conduction has forfeit an opportunity for partial conversion to work. The rate of loss of potential work is given in Availability notation as $T_0 s'$ [W], in which T_0 is the lowest temperature at which q' could realistically be rejected. In the present context $T_0 = T_C$. Substituting gives $s' = T_0 q'\{1/T_C - 1/T_E\}$:

$$W'_{lost} = T_0 s' = T_C q'\{1/T_C - 1/T_E\}$$

With T_E/T_C abbreviated to N_T:

$$= q'\{1 - 1/N_T\} \tag{11.5}$$

If the heat flow path of Figure 11.6 happens to be a Stirling engine, stationary but at operating temperature, then W'_{lost} is $q'\eta_C$, where η_C is Carnot efficiency between T_E and T_C – in other words, work *potentially* available has been lost at a rate equal *precisely* to that at which an *ideal* engine would have converted q'.

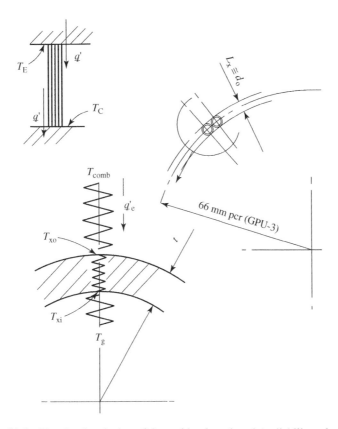

Figure 11.6   Notation for algebra of thermal bottleneck and Availability calculations

So where is the relevance to *real* engines? Under the heading *'Air pre-heaters'* Hagreaves (1991, at p. 203) puts flame temperatures at > 1800 °C and tube temperatures at 700 °C. Downstream of the heater-tube cage, temperature of combustion products has fallen to 800–1100 °C. To achieve significant heat transfer rate *inside* the individual expansion exchanger tube, gas temperature averaged over a cycle must be below nominal tube temperature. Tests of a resurrected GPU-3 engine were controlled in such a way that thermocouple probes installed inside 4 of the 40 heater tubes indicated 1200 or 1300 °F (650 and 700 °C respectively) according to fuel flow rate. The thermocouple is essentially a DC transducer so these temperatures were some sort of average. The probes would have responded to radiation from the inside of the tubes, suggesting an over-indication. With little else to go on, an 'educated guess' in the style of the legendary, late Professor G (Joe)Walker suggests mean temperature difference between heat source and point of heat reception by the gas can be 1800 °C–600 °C = 1200 °C! It puts the principal thermal bottleneck at the expansion end. The GPU-3 tests noted compression-space temperatures of 247 °F, or 119.4 °C – a penalty to be reckoned with (but *after* the main culprit has been addressed). The lower-most image of Figure 11.6 indicates schematically the thermal conduction path between combustion reaction products at T_{comb} and the working gas at the notional cycle mean T_{gxe} of its fluctuating temperature. The path may be seen as

Table 11.1   Extract from specification of General Motors' GPU-3 engine

---

8.9 kW brake power and 22.5% brake thermal efficiency at 3600 *rpm* when H_2-charged to 68 bar.

Expansion exchanger: 40 tubes $d_i = 3.02$ mm (internal) $\times d_o = 4.83$ mm (external).
Internal hydraulic radius $r_h = d_i/4 = 0.755$ mm.
Total flow path length L_x per (straight) tube = 245.3 mm.

Cross-section through tube cage intersects 80 tubes equi-spaced on 132 mm pitch circle dia.
   (40 tubes cylinder side, 40 regenerator side).

Combustion-side gas path length = d_o of individual tube, that is, 4.83 mm.
Inter-tube gap in peripheral direction: 0.353 mm.
Hydraulic radius: 0.443 mm.

---

three 'thermal bottlenecks' in series: (a) between products at T_{comb} and external surface of the expansion exchanger tubes at T_{xo}, (b) between T_{xo} and the inner surface of the tubes at T_{xi}, (c) between T_{xi} and working fluid at T_{gxe}.

Figures in Table 11.1 are for General Motors' GPU-3 engine at the peak power point. According to the thinking which led to Equation 11.5, rate of loss of available work between T_{comb} and T_{gxe} is $T_C q'\{1/T_{comb} - 1/T_{gxe}\}$. Substituting numerical values for T_{gxe}, and so on from Table 11.1 suggests:

$$W'_{lost}/q'_E = 310\{1/2023 - 1/923\}$$

$$= 0.1826 - \text{that is, } W'_{lost} > 18.26\% \text{ of } q'_E!$$

The penalty of 18% or so is incurred *before heat even enters the engine*! As a percentage of the shaft output of the 22.5% efficient GPU-3, the loss is 81%!! Moreover, it can be argued this is an *under-estimate*: in reality, heat flow reverses cyclically, with peak values in excess of the mean q'_E. (Entropy is created – and available work lost – in both directions of heat flow.)

Some may find the value implausibly large. To confirm that this is not a quirk of the Availability approach, the identical value is achieved by traditional algebra in the final section of this Chapter.

The focus to this point has been on the high-temperature end of the cycle. It should nevertheless be obvious that achieving high efficiency requires that *the working fluid* (rather than the heat exchangers) should receive and reject heat at temperatures as close as possible to those of source and sink respectively.

The criterion is consistent with minimizing temperature differences along the heat-flow path. In engineering terms, cutting down on such losses amounts to minimizing *'thermal bottlenecks'*.

Figure 11.6 identifies three bottlenecks (or 'thermal impedances') in series at the expansion end alone:

Finite thermal conductivity k [W/mK] of tube walls. The 8.9 kW output of the GPU-3 at 22.5% efficiency equates to a heat input rate of 8.9/0.225 = 39.5 kW. Steady conduction through the walls of the 40 exchanger tubes is governed by the equation $q' = -kA\Delta T/t_w$, where k is coefficient of thermal conductivity [W/mK], A is total surface area [m^2] of the

40 tubes, ΔT [°C] is the difference in temperature between internal and external surfaces, while t_w is radial thickness [m] of the tube wall. k for AISI 316 stainless steel at 600 °C is 20 W/mK. Making use of data from Table 11.1:

$$\Delta T_{cond} = \frac{39.5E + 03[W] \times 0.905E - 03[m]}{20\,[W/mK] \times 40 \times 0.245\,[m] \times \pi \times 4.83E - 03[m]}$$

$$= 12.02\,°C$$

Availability Theory sees heat flow across ΔT_{cond} as responsible for *loss of available work*.

ΔT has been used (rather than ε_T) because, in the context, ΔT does not have the sense of a perturbation error.

Restricted NTU between combustion products and outer tube surface. NTU is defined as the product of Stanton number St with the ratio of path length L to hydraulic radius r_h, viz, $NTU = StL/r_h$. On the working gas side $L/r_h = 245.3/0.755 = 325$. On the combustion side $L/r_h = 4.83/0.443 = 10.9$. For lack of data specific to elevated temperatures, Kays' and London's correlations $(StPr^{2/3} - Re)$ are used: At $Re = 10\,000$, St on the working fluid side (internal to cylindrical tube) ≈ 0.003. On the combustion side, St perpendicular to a tube bank ≈ 0.008. Radiation is known to enhance values based on convection alone, suggesting an effective value somewhat higher. St is inflated (arbitrarily) by 50% to $St \approx 0.012$ to reflect this. Heat flux is common to both sides, allowing the ratio $\Delta T_{ext}/\Delta T_i$ of $(325 \times 0.003)/(10.9 \times 0.012) = 7.45$ [–]. For a total temperature drop of 1200 °C between T_{comb} and T_{gxe} this converts to $\underline{\Delta T_i} \approx 140\,°C$. Taking account of ΔT_{cond} from the earlier conduction calculation, $\Delta T_{ext} \approx 1050\,°C$.

The independent approach of Chapter 3 suggested $\underline{\Delta T_i} \approx 150\,°C$. Could it be that these back-of-envelope values have come within striking distance of reality?

The massive loss of potentially available work attributable to the thermal bottleneck is the incentive for exploring the unorthodox exchanger geometry recounted in Chapter 14.

11.5   Available work lost – conventional arithmetic

The loss is equal to the difference between W' for the 'perfect' engine operating between ultimate limits T_{comb} and T_C and W' and the same engine operating from an upper temperature limit reduced by the ΔT over which heat is conducted from T_{comb} to the temperature of heat reception by the working gas:

$$W'_{Tcomb} = \underline{q}'_E\{1 - 310/2023\} = 0.8467\underline{q}'_E$$

To isolate the high-temperature loss, counterpart loss at the compression end is suppressed by setting T_{gxc} constant and equal to T_C. The resulting cycle remains an idealization, but this time it operates from more realistic temperature T_{gxe}.

W' is now:

$$W'_{Tgxe} = \underline{q}'_E\{1 - T_C/T_{gxe}\} = \underline{q}'_E\{1 - 310/923\} = 0.6641\underline{q}'_E$$

The difference $W'_{Tcomb} - W'_{Tgxe}$ is $\underline{q}'_E(0.847 - 0.664) = 0.1826\underline{q}'_E$ – *precisely the loss previously calculated explicitly in a single step using Equation 11.5.*

12

FastTrack gas path design

12.1 Introduction

The chapter title reflects an overriding aim: to eliminate – or at least to streamline – one or more of the many design challenges lying between concept (power, *rpm*, working fluid, etc.) and eventual prototype drawings.

If you can live with the idea of an engine whose gas process cycle precisely mimics that of one of the benchmark engines (GPU-3, P-40, etc.) then *FastTrack* may be the design process for you!

Consistent with priorities, a complete gas path design is illustrated first, derivation and justification offered later. If you want to skip the latter, it is sufficient to note that the analysis behind the design charts – mostly parallel-scale alignment nomograms – achieves, in a derivative design using air/N_2 or H_2, a replication of the gas processes of General Motors' GPU-3 at its peak published power point of 8.95 kW at 3600 *rpm*. When charged with He the GPU-3 delivered 4.25 kW at 2500 *rpm*. Scaling from this performance is via the same computational sequence, but the graphical short-cut calls for a different set of charts supplied here under Section 12.6 – *Alternative start point*.

It is likely that the GPU-3 was not optimized for either set of operating conditions. If this is the case, then the derivative design is also not optimum. It is therefore capable, in principle, of performance improvement – but not, obviously, by the present method. In partial compensation, peak power point Beale number N_B for the GPU-3 is an impressive 0.18: there are worse designs from which to scale!

Gas process replication is achieved *as closely as anyone could reasonably hope to calculate.* Nevertheless, a design methodology is not a design methodology until tried and tested in practice. *FastTrack* has yet to be applied to that phase. The design and build of a power-producing engine is an avid consumer of time and money. Anyone proposing to convert a gas path design into hardware in the meantime should *at the very least* satisfy him/herself that the background offered by Section 12.5 *Rationale* provides the assurance required by the proposed financial commitment.

The *Fast-Track* sequence eliminates the computer. Internal diameters d_x, lengths L_x, and numbers n_{Tx} of tubes are read in an instant from high-resolution charts. Use of the hand-calculator can kept to a minimum.

Stirling Cycle Engines: Inner Workings and Design, First Edition. Allan J Organ.
© 2014 John Wiley & Sons, Ltd. Published 2014 by John Wiley & Sons, Ltd.

12.2   Scope

A single cylinder-set (or gas path) is considered, as in the coaxial 'beta', or the V-alpha configuration. The Rider concept and the additional dead space of the 'gamma' configuration are not compatible with the arithmetic. Moreover, results may be misleading if applied to one cylinder-set (one-quarter) of a four-cylinder 'Rinia' machine, whose first harmonic of the volume phase angle, being 90 degrees, is in conflict with the constraints. These are:

Temperature ratio $N_T = T_E/T_C$:   3.3 (e.g., $T_E = 1020$ K, $T_C = 310$ K)
Expansion and compression exchangers both of multi-tube type, with tubes of internal diameter d_x, lengths L_x and respective number n_{Tx}.
Coaxial, uniform-bore machine requires prototype (GPU-3) values for kinematic phase angle β and piston/displacer displacement ratio λ:

$$\beta = 60 \text{ degrees}; \qquad \lambda = 0.98$$

Achieving these values in the opposed-piston or 'V' configuration calls for:

thermodynamic volume ratio κ: 1.0
thermodynamic phase angle α: 120 degrees

α *and* κ *have been converted via the formulae for simple-harmonic (SHM) volume variations* (e.g., those derived by Finkelstein, 1960a). Those of the GPU-3 are not SHM!

Regenerator: stacked from wire screens of hydraulic radius r_h determined by wire diameter d_w and mesh number m_w. Volume porosity $\P_v$ is pre-set to 0.75. Resulting Beale number (given adequate heating and cooling provision): up to 0.18 depending on mechanical efficiency.

12.3   Numerical example

The example proceeds with air or N_2 as working fluid – and by addressing the *rpm* requirement: *rpm* is the design parameter which frequently offers the least flexibility, being the value demanded by an electric alternator or by the screw of a boat. If it is a priority to avoid gearing, then design is almost certainly driven by *rpm*. This example uses *rpm* = 1500.

All other things being equal, air/N_2 tend to demand a larger number of exchanger tubes than H_2 (or He). One superfluous tube is one extra potential leak. To this extent, the start point is the target *rpm* and tolerable number n_{Tx} of the exchanger tubes.

Figures 12.1 to 12.7 allow design for either H_2, by using the right-side graduations of double-sided scales, or air/N_2 by using the left-side graduations. (The different isentropic index γ of He calls for an alternative set of charts.) Number of expansion exchanger tubes n_{Txe} for the air/N_2-charged derivative is located at the middle scale of Figure 12.1 using the left-side graduations. A straight-edge pivoted against acceptable n_{Txe} and intersecting target *rpm* on the right-most scale shows achievable power [W] on the left-most scale. 1500 *rpm* and 43 tubes furnish the example – in this case because they lead to a convenient 1000 W (1 kW) on the power scale.

Swept volume V_{sw} is read in convenience units (cc) from Figure 12.2: a straight edge or pencil line through 1000 W on the left-most scale, and through 1500 *rpm* on the right-most scale gives 32.65 cc at the appropriate (left-side) graduations of the centre scale. Delay

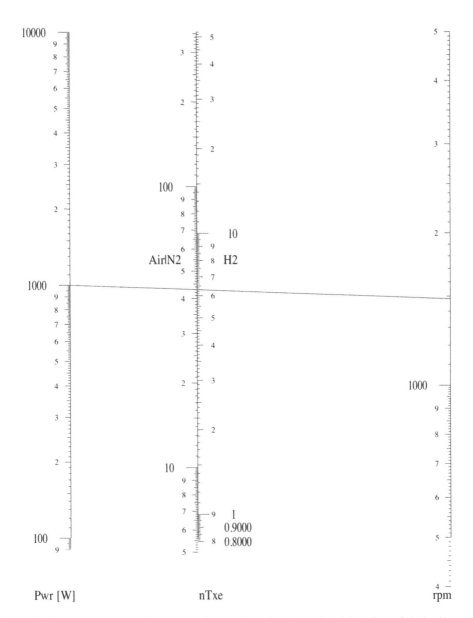

Figure 12.1 Locate power [W] and *rpm*, join with straight line. For air/N₂-charged derivative read number of tubes n_{Txe} for expansion exchanger from left-side graduations of centre (n_{Tx}) scale. For H₂-charged derivative read from right-side graduations

rounding this (to, say, 32.5 cc) because it can make more sense to see what V_{sw} results from the use of rational values of bore and stroke.

Now for the reckoning! On the left-most (power) scale of Figure 12.3 connect 1000 W with 1500 *rpm* on the *rpm* scale and read 67.23 bar from the left-side graduations of the pressure (p_{ref}) scale. This may be the point for a re-think: hydrocarbon lubricant can react explosively

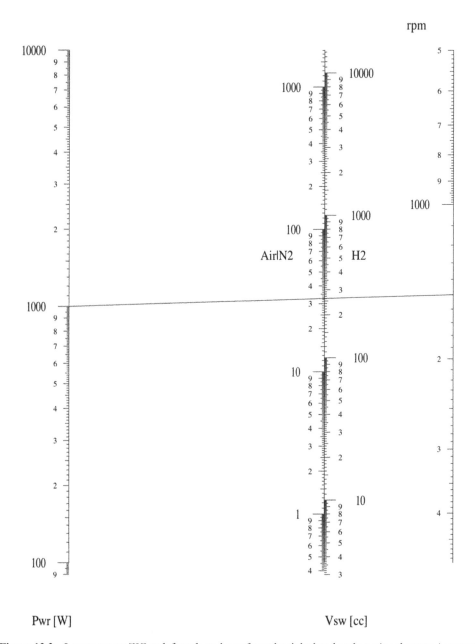

Figure 12.2   Locate power [W] on left scale and *rpm* from the right-hand scale *noting that* rpm *increase downwards*. Join with straight line and note intersection with V_{sw} scale, reading from the graduations on the left if chosen working fluid is to be air/N$_2$, those to the right if H$_2$

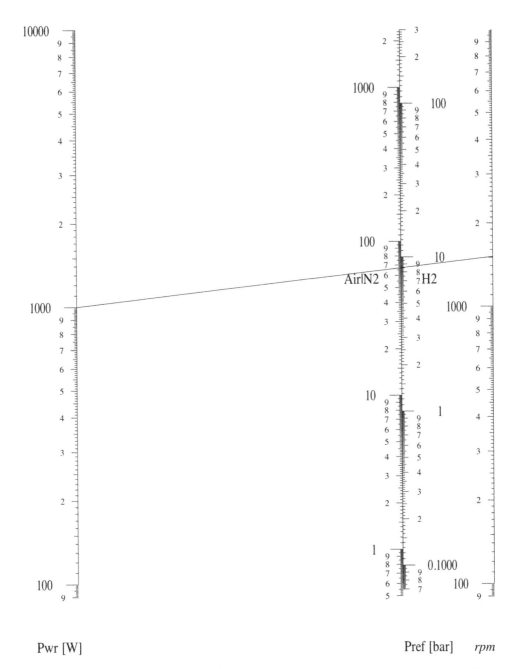

Pwr [W]   Pref [bar]   *rpm*

Figure 12.3   Locate power [W] and *rpm*, join with straight line, and read charge pressure p_{ref} [bar] from left side of centre scale for air/N_2-charged design, right side of same scale for H_2-charging

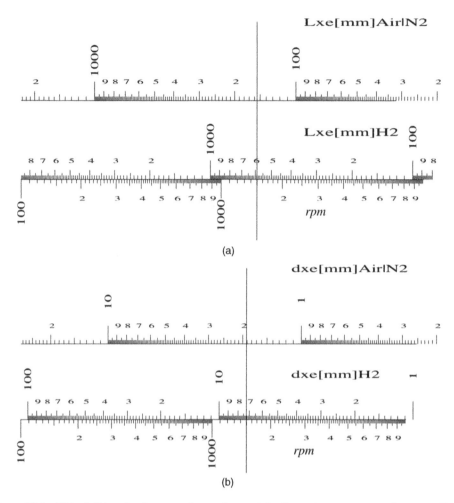

Figure 12.4 Virtual slide-rules for expansion exchanger tube dimensions versus *rpm* (lowermost horizontal scale) for air/N_2 (uppermost horizontal scale) as derivative gas. For H_2 as working gas for derivative design read middle horizontal scale against cursor. (a) Flow passage length L_{xe} [mm] (b) Internal diameter d_{xe} of individual tube [mm]

with air – and the higher the pressure the greater the hazard. Options include: (1) proceeding, but with N_2 rather than air (2) pressing ahead in anticipation of completing the mechanical design with the aid of 'solid oil' bearing technology and polymer-alloy rubbing seals throughout, (3) returning to square-one and targeting reduced power, and (4) turning to He as working fluid. In case (4) it will be necessary to proceed to charts for He, which can be found after Section 12.5 *Rationale* below.

Flow passage length L_{xe} of the individual expansion exchanger tube is read in convenience units (mm) from the air/N_2 scale of the 'virtual slide-rule' of Figure 12.4: a vertical line (the 'cursor') through 1500 *rpm* of the bottom (*rpm*) scale cuts the air/N_2 scale at 155 mm. A value of $d_{xe} = 1.91$ mm is obtained from Figure 12.5 in the same fashion – and completes

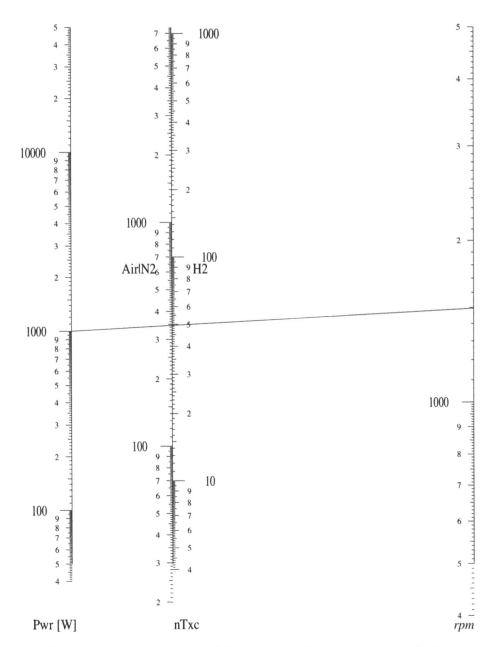

Figure 12.5 Locate power [W] and *rpm*, join with straight line, and read number of tubes n_{Txc} for compression exchanger from left side of centre (n_{Tx}) scale for air/N$_2$-charged derivative, right side of same scale for H$_2$-charging

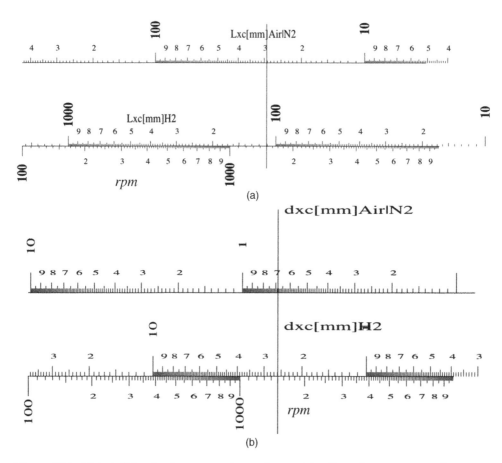

Figure 12.6 Virtual slide-rules for compression exchanger tube dimensions versus *rpm* (lowermost horizontal scale) for air/N_2 (uppermost horizontal scale) as derivative gas. For H_2 as working gas for derivative design read middle horizontal scale against cursor. (a) Flow passage length L_{xc} [mm] (b) Internal diameter d_{xc} of individual tube [mm]

the internal design of the expansion exchanger. Note the advantageous partial de-coupling from *external* heat transfer design: *internal* design is essentially independent of tube spacing, whereas that very spacing determines the hydraulic radius 'seen' by the combustion products flowing between the tubes.

Resist the temptation to round the d_{xe} value of 1.91 mm to 1.9 (or 2.0 mm) pending consultation of the stock lists of tube suppliers/manufacturers (*http://www.welleng.co.uk/tube.html*). Sizes which are irrational at first sight may arise on converting from imperial inch or the result of coincidental combination of a rational outside diameter with a wall thickness specified to a US or UK standard gauge specification. *Not one* of the internal diameters on the Fractional Metric stock list of Coopers Needleworks (*http://www.finestainlesstube.com/*) is a rational size! In the 'thin-wall' range a 14-gauge tube is listed having internal diameter of 1.828 mm.

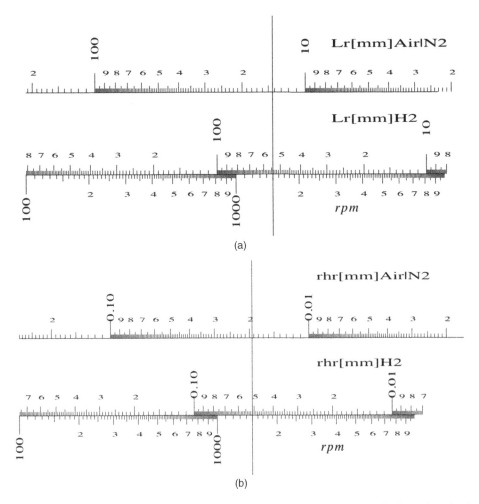

Figure 12.7 Virtual slide-rules for matrix parameters versus *rpm* (lowermost horizontal scale) for air/N_2 (uppermost horizontal scale) as derivative gas. For H_2 as working gas for derivative design read middle horizontal scale against cursor. (a) Regenerator stack height L_r [mm] (b) regenerator hydraulic radius r_{hr} [mm]

Time for a check: whatever happened to Beale number N_B in these deliberations?! Using the p_{ref}, and so on from the charts, and remembering to convert from convenience units to a consistent system:

$$N_B = 1000\,[\text{W}]/\{67.23\text{E}+05\,[\text{Pa}]\times 32.87\text{E}-06[\text{m}^3]\times 1500/60[-]\}$$

$$= 0.181(\text{cf the } 0.18 \text{ anticipated at the outset})$$

Compression exchanger design proceeds in parallel fashion from a chart-set reduced in number by two, reflecting the fact that values for p_{ref} and V_{sw} do not need to be acquired a second time. The charts follow.

Regenerator design is completed as for the tubular exchangers – except for a step to convert hydraulic radius r_{hr} via volume porosity $\P_v$ to wire diameter d_w and mesh number m_w which specify the required grade of wire gauze.

Design charts follow for hydraulic radius r_{hr} and overall stack height L_r. A simple exercise for the hand calculator converts hydraulic radius r_{hr} to wire diameter d_w [mm] mesh number m_w [wires/mm or wires/inch]. In the equation, symbol $\P_v$ represents volume porosity, which has been pre-set for this example to 0.75 [–].

$$d_w = 4r_{hr}[\text{mm}]\,(1 - \P_v)/\P_v[\text{mm}] = 0.0253 \text{ mm, or } 0.001 \text{ inch.}$$

$$m_w = \P v/\pi r_{hr}\,[\text{wire/mm}] = 12.49 \text{ wire/mm, or } 317 \text{ wire/inch } (\approx 320 \text{ mesh}).$$

A manufacturer of precision wire gauzes holding a comprehensive stock range G Bopp and Co. Ltd. (*http://www.boppmesh.co.uk/*)

12.4   Interim comment

The specification derived from the peak power of the H_2-charged GPU-3 is eminently viable in terms of number, length and diameter of exchange tubes. On the other hand, its charge pressure p_{ref} of 67.23 bar is not for the faint-hearted. So it is worth reflecting that the outcome does not define prospects for the 1 kW air-charged engine – whether at 1500 *rpm* or otherwise:

- Scaling is but one way of advancing from performance target (power, *rpm*) to gas path specification. Of two known scaling options *FastTrack* is the arguably the more rigorous – but correspondingly the less flexible.
- It is inconceivable that the GPU-3 could not be re-designed so as to achieve rated output (8.95 kW at 3600 *rpm*) from reduced p_{ref} and proportionally increased V_{sw}. Corresponding to each such hypothetical re-design there is an (hypothetical) air-charged design derived by scaling to the present requirements – 1 kW at 1500 *rpm*. In each such case p_{ref} will be reduced relative to the troublesome 67.23 bar.
- Unless the GPU-3 happens to be an absolute optimum, the scaled derivative is unlikely to be the sole design capable of target power (1 kW) on air/N_2.
- Pending verification of these possibilities it will be of interest to look at a helium-based derivative design achieved by scaling from the performance of the He-charged GPU-3. The outcome will be more meaningful against an account of the mechanism underlying the scaling sequence.

12.5   Rationale behind *FastTrack*

Like every other feature of this text, *FastTrack* is based in *Similarity principles*. This is **not** the same thing as insisting that all Stirling engines are rigorously similar (although a degree of intrinsic similarity has been argued). Rather, it is a recognition of the benefits of studying, say, the buckling characteristics of structural columns in terms of the dimensionless parameter *slenderness ratio d/L*, rather than dealing separately with each *d* and each *L*.

The gas processes of two engines charged to different pressures p_{ref} and/or running at different *rpm* can be arranged to be similar only if the engines share identical numerical values of a limited set of basic parameters:

The ***basic similarity parameters*** of the Stirling engine are:

Temperature ratio $N_T = T_E/T_C$
Volume ratio $\kappa = V_C/V_E$
thermodynamic phase angle α (or, for practical purposes, the first harmonic thereof)
Dead space ratios $\delta = V_d/V_w$: $\delta_{xe} = V_{dxe}/V_{sw}$, $\delta_r = V_{dr}/V_{sw}$, and so on.
Specific heat ratio or isentropic index γ.

Achieving similarity in practice calls for further constraints:

Geometric similarity of the flow passages: in both cases, cylindrical tubes, or slots of the same depth/width ratio; regenerators of geometrically similar matrix material, e.g., square-weave wire gauze of the same volume porosity $\P_v$.

It is now possible (in principle) to constrain the operating conditions of the two machines (working gas, reference pressure p_{ref}, *rpm*) and the finer detail of the gas path (internal diameters d_x, lengths L_x and respective numbers n_{Tx} of tubes) so as to ensure that (variable) temperature distributions and pressure drops follow identical histories over the entire 360 degree cycle. This results automatically by arranging Reynolds number Re and Mach number Ma at any given location to undergo the same respective variations over a cycle. If the cycle histories of local Re are the same, so are the cycle histories of Stanton number St and of friction factor C_f at comparable locations. If, simultaneously with this, the Ma histories are the same, then local cycle histories of fractional pressure drop dp/p may be arranged to be identical also.

Analytically complex? With the *basic* similarity conditions already satisfied, less daunting than might be feared!

The Reynolds number is defined as $Re = 4\varrho u r_h/\mu$. It may be re-expressed in terms of mass rate $m' = dm/dt$ on noting that $m' = \varrho u A_{ff}$, in which A_{ff} is free-flow area. A_{ff} for the multi-tube exchanger is $\frac{1}{4} n_{Tx} \pi d_x{}^2 L_x$.

An earlier chapter *Intrinsic Similarity* introduced *specific mass* $\sigma = mRT_C/p_{ref}V_{sw}$ and specific mass rate $\sigma' = m'RT_C/\omega p_{ref}V_{sw}$. Substituting the expansion for A_{ff} and the definition of σ' into Re:

$$4Re/\pi = \frac{\sigma'}{(d_x/V_{sw}{}^{1/3})n_{Tx}} \frac{p_{ref}}{\omega\mu_0 f(T)} \frac{\omega^2 V_{sw}{}^{2/3}}{RT_C}$$

The term $f(T)$ corrects a datum value μ_0 of coefficient of dynamic viscosity μ to the value at the nominal temperature (T_E or T_C) of the exchanger, for example, $\mu_{TE} = \mu_0 f(T_E)$.

It has been demonstrated in an earlier chapter that the cycle history of specific mass rate σ' computed from the ideal adiabatic reference cycle at any reproducible location varies little between the benchmark engines GPU-3, P-40, and so on. Where hypothetic engines have numerical values of the basic similarity parameters in common, an *infinite range* of such engines has *identical* cycle history of σ'.

Term $\omega^2 V_{sw}{}^{2/3}/RT_C$ is equivalent to Mach number, Ma^2, and requires to be independently similar, as σ' already is. For the cylindrical duct $r_h = d_x/4$. Angular speed ω can be re-written

$\omega = 2\pi rpm/60$. Similarity of Re (and thus of St and C_f) is now achieved by ensuring commonality of the numerical value of the dimensionless group DG_{Re}:

$$DG_{Re} = p_{ref}/\{rpm\mu_0 f(T)(d_x/V_{sw}^{1/3})n_{Tx}\}$$

Pressure drop dp in steady flow in a parallel duct of hydraulic radius r_h and length L_x is: $dp = \frac{1}{2}\rho u^2 C_f L_x/r_h$. Using $r_h = d_x/4$ again, fractional pressure drop dp/p over length L_x of the exchanger is:

$$dp/p = 2\gamma [u^2/(\gamma RT)]C_f(Re)L_x/d_x$$

The term $u^2/(\gamma RT)$ is the square of Mach number Ma. With similarity of C_f taken care of (by similarity of Re), similarity of dp/p is ensured by arranging similarity of the product $\gamma Ma^2 L_x/d_x$. The easiest way of achieving this is to impose similarity of the groups independently. γ is already subject to a basic similarity condition. Ma is dealt with by making the same substitutions as for Re. There are now the similarity groups DG_{Ma} and DG_{Ld}:

$$DG_{Ma} = rpmV_{sw}/\{(\sqrt{RT_C})d_x^2 n_{Tx}\}$$

$$DG_{Ld} = L_x/d_x$$

A plausible design requirement is a specified value of power Pwr [W] at specified rpm. This calls for the designer to establish the V_{sw}, p_{ref} and, for each exchanger, the L_x, d_x, and n_{Tx} – five numerical values – which will cause the engine to achieve the specified performance. So far there are three similarity conditions DG_{Re}, DG_{Ma}, and DG_{Ld}, so two more are required.

The first is acquired by re-expressing – in terms of L_x, d_x, and n_{Tx} – similarity of exchanger fractional dead space $\delta_{xe} = V_{dxe}/V_{sw} = \frac{1}{4}\pi d_x^2 L_x n_{Tx}/V_{sw}$. The constants may be dropped, yielding dead-space criterion $DG_{\delta x}$:

$$DG_{\delta x} = d_x^2 L_x n_{Tx}/V_{sw}$$

Finally, look no further than the original similarity parameter N_B variously attributed to Finkelstein and Beale. The constant which converts rpm to frequency f is dropped, and convenience units used for p_{ref} and V_{sw}. The symbol N_B is replaced by DG_{NB} (because the numerical value will no longer be ≈ 0.15):

$$DG_{NB} = power [W]/\{p_{ref} [bar]V_{sw}[cc] rpm\}$$

The (unknown) parameter values for the engine to be designed (the *derivative*) can now be equated to corresponding values for an engine of known specification and (preferably exemplary) performance – the *prototype*. The process transfers to the right-hand side values already prescribed for the derivative – power Pwr [W] and rpm. Variables for the derivative are on the left:

$$p_{ref}/\{(d_x/V_{sw}^{1/3})n_{Tx}\} = DG_{Re} \; rpm \; \mu_0 f(T)$$

$$V_{sw}/\{d_x^2 n_{Tx}\} = DG_{Ma}/rpm\sqrt{RT_C}$$

$$L_x/d_x = DG_{Ld}$$

$$d_x^2 L_x n_{Tx}/V_{sw} = DG_{\delta x}$$

$$p_{ref} V_{sw} = Pwr/(DG_{NB} \; rpm)$$

The equations are non-linear in the unknowns. An apt succession of division of rows leads to explicit expressions for individual solutions. On the other hand, a solution which serves as a prototype for a more comprehensive treatment is probably a worthwhile investment. The equations are accordingly linearized by taking logs. The resulting array of coefficients is:

p_{ref}	V_{sw}	L_x	d_x	n_{Tx}	RHS
1	⅓	0	−1	−1	$\log\{DG_{Re}\, rpm\, \mu_0 f(T)\}$
0	1	0	−2	−1	$\log\{\sqrt{RT_C}DG_{Ma}/rpm\}$
0	0	1	−1	0	$\log(DG_{Ld})$
0	−1	1	2	1	$\log(DG_{\delta x})$
1	1	0	0	0	$\log\{Pwr/(DG_{NB}\, rpm)\}$

This is easily coded for numerical solution (by a library routine such as SIMQX). Five *DG* are substituted by respective prototype values. For a given prototype, one solution run serves for all derivative designs.

For reasons which a competent analyst might have anticipated, solutions for p_{ref}, V_{sw} and n_{Tx} emerge in the form $p_{ref} = C_1 Pwr^{a1}\, rpm^{b1}$, $V_{sw} = C_2 Pwr^{a2}\, rpm^{b2}$, $n_{Tx} = C_3 Pwr^{a3}\, rpm^{b3}$. Both L_x and d_x are unexpectedly independent of *Pwr*, and so can be represented $L_x = C_4 rpm^4$ and $d_x = C_5 rpm^5$. This allows the former three solutions to be graphed against *Pwr* with *rpm* as parameter. Each of the latter is evidently a unique curve plotted against *rpm*. The parallel-scale alignment nomogram is equivalent to the conventional *x-y-parameter* presentation. Its superior resolution explains the choice of format for the design charts of this chapter.

12.6 Alternative start point – GPU-3 charged with He

Charged with helium to maximum rated pressure (69 bar abs), brake power output was 4.25 kW at 2500 *rpm*. This provides a performance point for scaling to derivative designs based on working fluids having isentropic index γ equal to that of He, namely 1.66. Beale number N_B for these conditions is:

$$N_B = 4250\,[\text{W}]/\{69.0\text{E}+05\,[\text{Pa}] \times 118.63\text{E}-06[\text{m}^3] \times 2500/60\,[-]\}$$

$$= 0.1246$$

The set of design charts which follows has been generated for the revised prototype performance point. The derivative design will again target 1 kW at 1000 *rpm*. The design sequence precisely replicates that already illustrated for the study based on air/N_2.

From the chart of Figure 12.8 the number of expansion tubes n_{Txe} is 12. From the chart of Figure 12.9 V_{sw} is now 169.67 cc. From Figure 12.10 the charge pressure p_{ref} is the eminently manageable 18.91 bar abs (17.91 bar gauge).

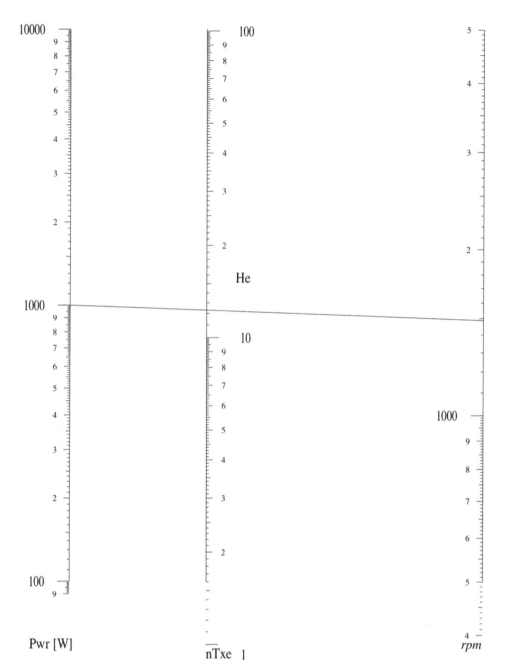

Figure 12.8   Chart corresponding to Figure 12.1. To be used only for scaling to He-charged derivative from GPU-3 performance point on He

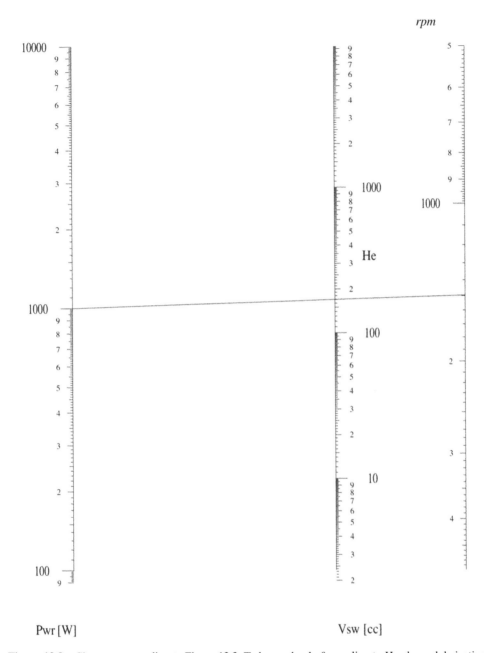

Figure 12.9   Chart corresponding to Figure 12.2. To be used only for scaling to He-charged derivative from He-charged GPU-3 performance

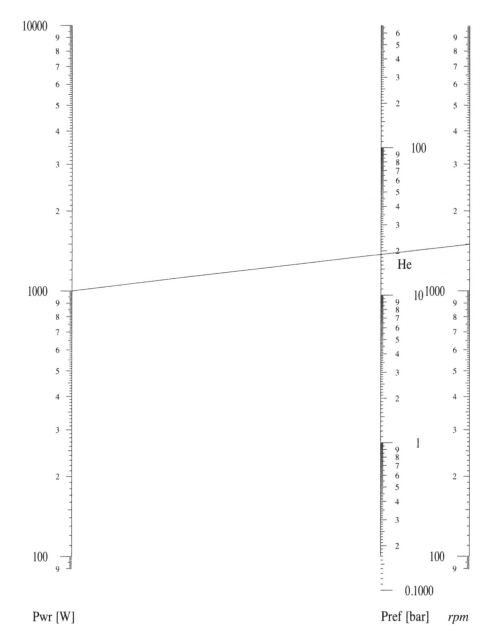

Pwr [W]   Pref [bar]   *rpm*

Figure 12.10   Chart corresponding to Figure 12.3. To be used only for scaling to He-charged derivative from He-charged GPU-3 performance

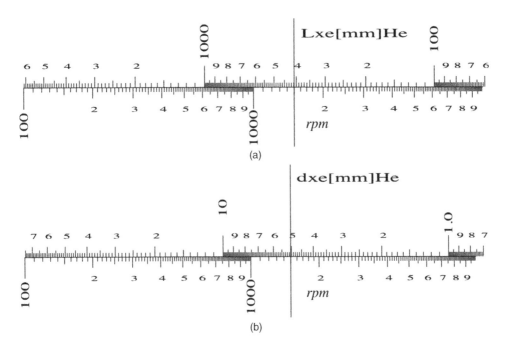

Figure 12.11 Virtual slide-rules for expansion exchanger tube dimensions versus *rpm* (lowermost horizontal scale) for He as derivative gas. (a) Flow passage length L_{xe} [mm] (b) Internal diameter d_{xe} of individual tube [mm]

The Beale number check is:

$$N_B = 1000\,[\text{W}]/\{18.91\text{E} + 05\,[\text{Pa}] \times 169.67\text{E} - 06[\text{m}^3] \times 1500/60[-]\}$$

$$= 0.1246 - \text{precisely the value calculated for prototype conditions}$$

The cursor of the virtual slide-rule of Figure 12.11(a) is already set at 1500 *rpm*, and intersects the L_{xe} scale at 408.83 mm. An expansion exchanger cage modelled after that of the GPU-3 would be uncharacteristically tall at over 200 mm. Should this prove unacceptable there may be an interesting application for the helical-spiral tube concept dealt with in Chapter 11 and already specified for the *mRT*-1k. The small number of tubes n_{Txe} perfectly complements this option.

Number n_{Txc} of compression tubes (Figure 12.12) is 96. From the virtual slide-rules of Figure 12.13 length L_{xc} and internal diameter d_{xc} are 59.17 and 1.80 mm respectively.

Figure 12.14 gives regenerator stack length L_r of 37.67 mm and hydraulic radius r_{hr} of 0.051 mm. Volume porosity $\P_v$ remains pre-set at 0.75 [−]:

$$d_w = 4r_{hr}\,[\text{mm}]\,(1 - \P_v)/\P_v\,[\text{mm}] = 0.068\,\text{mm, or } 0.00268\,\text{inch.}$$

$$m_w = \P_v/\pi r_{hr}\,[\text{wire/mm}] = 4.68\,\text{wire/mm, or } 119\,\text{wire/inch}\,(\approx 120\,\text{mesh}).$$

Assuming mesh thickness equal to $2d_w$, number of gauzes is approximately $L_r/2d_w = 277$. It is easier to punch discs from coarse mesh than from fine, so a relatively cheap regenerator is in prospect.

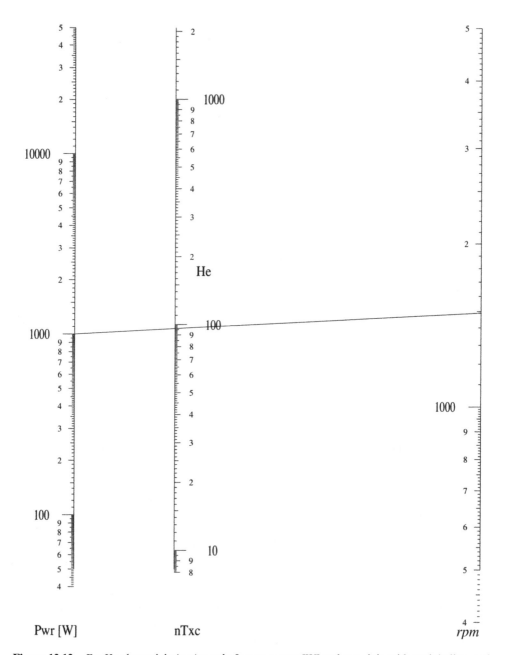

Pwr [W]   nTxc   *rpm*

Figure 12.12 *For He-charged derivative only.* Locate power [W] and *rpm*, join with straight line, and read number of tubes n_{Txc} for compression exchanger from left side of centre (n_{Tx}) scale

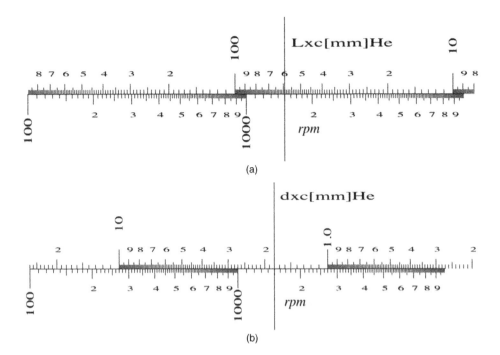

Figure 12.13 (With He as gas for derivative design.) Virtual slide-rules for compression exchanger tube dimensions versus *rpm*. (a) Flow passage length L_{xc} [mm] (b) Internal diameter d_{xc} of individual tube [mm]

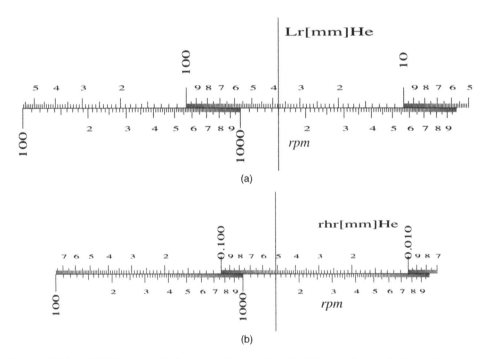

Figure 12.14 (With He as gas for derivative design.) Virtual slide-rules for matrix dimensions versus *rpm*. (a) Regenerator stack height L_r [mm] (b) regenerator hydraulic radius r_{hr} [mm]

13

FlexiScale

13.1 *FlexiScale?*

Scaling by *FastTrack* from the USS P-40, for example (rather than from the GPU-3) would call for an alternative set of charts. The single chart set of *FlexiScale* accepts as prototype any engine for which the designer has access to a specification.

The chart scales are dimensionless ratios – rpm^*/rpm, d_{xe}^*/d_{xe}, and so on – rather than the 'convenience' units employed for *FastTrack*. (Asterisk * always denotes the derivative.) Such a ratio has the same numerical value in any consistent system of units – mm/mm, in/in, ft/ft, and so on.

Chart display and serviceability benefit from recognition that there are only three gases of practical interest: hydrogen (H_2), helium (He) and nitrogen[1] (N_2). This allows the appropriate scales to be marked in terms of the gases themselves (H_2 to N_2, He to He, etc.), rather than graduated with a continuously varying ratio of gas properties.

The 'flex' of *FlexiScale* comes at minor cost:

- A modest exercise in simple division is required to give entry point(s) for each chart.
- A value of *rpm* for the derivative design requires to be set at the outset. Most practical applications require power unit to be matched to application. If the latter calls for direct drive, then *rpm* are pre-determined, and the designer has no choice anyway.

An example precedes exposition of the underlying rationale. Re-using the scaling case of the previous chapter will provide a degree of cross-checking. The prototype is therefore the GPU-3 at its H_2 operating point (Table 13.1 below). As previously, scaling is to an air-charged derivative at 1500 *rpm*. *Flexi-Scale* provides for target power to be left open at this stage.

---

[1]For the purposes of design calculations N_2 and air are interchangeable. However, N_2 is inert for present purposes, whereas air can react dangerously with hydrocarbon lubricants above a certain combination of pressure with temperature.

Stirling Cycle Engines: Inner Workings and Design, First Edition. Allan J Organ.
© 2014 John Wiley & Sons, Ltd. Published 2014 by John Wiley & Sons, Ltd.

Table 13.1 Prototypes whose performance point has been achieved with working fluid having isentropic index $\gamma = 1.4$. Any one may be scaled to an air-charged or an H_2-charged derivative

		GPU-3 (H_2)	MP1002EQ[+]	P-40
Parameter values of prototype gas path which transfer directly to derivative:				
N_T		3.276	2.92	3.07
γ		1.4	1.4	1.4
κ_{eq}		~1.0	1.031	0.967
α_{eq} (deg)		120	120	90
$\P_v$		0.75	0.8	0.685

Values of dead space ratios δ_{xe}, δ_r, δ_{xc} transfer by default via scaled products typified by $(\frac{1}{4}\pi)d_{xe}^2 L_{xe} n_{Txe}/V_{sw}$ (see rest of Table below).

Raw prototype data having dimensions:				
Pwr	W	8950	250	11 250
rpm	1/min	3600	1500	4000
V_{sw}	cc	118.63	62	134
p_{ref}	bar	69	15	150
L_{xe}	mm	245.3	37.25	260.0
L_r	mm	22.6	28.0	44.0
L_{xc}	mm	46.1*	37.25	90.0
d_{xe}	mm	3.02	0.8[+]	3.0
r_{hr}	mm	0.0304	0.056	0.0273
d_{xc}	mm	1.08	0.8[+]	1.0
A_{ffr}	mm^2	2406	728	2×1936
n_{Txe}	–	40	303[+]	18
n_{Txc}	–	312	303[+]	400
R	J/kgK	4120	287	4120
μ_0	Pas	0.008E–03	0.017E–03	0.008E–03

[+]The MP1002EQ specification is that of the Philips MP100 2CA with the exception of expansion and compression slot sets, which have been substituted by tube sets having the same length, hydraulic radius, free-flow area and dead space as the original slots. For the conversion algebra, see Chapter 11.
*See note in text on numerical value of L_{xc}.

13.2   Flow path dimensions

The ratio of prototype to derivative *rpm*, denoted *rpm**/*rpm* on the chart of Figure 13.1, is $1500/3600 = 0.4167$ [–]. Scaling is from H_2 to air/N_2, that is, to the point labelled *H_2-to-Air* on the *gas**/*gas* scale. Connecting these two points with a straight line gives the ratio of exchanger passages lengths, L_x, derivative to prototype, viz L_x*/$L_x = 0.6334$ [–]. From Table 13.1 L_{xe} is 245.3 mm:

$$L_{xe}* = 245.3 \times 0.6334 = \mathbf{155.37 \ mm}$$

Table 13.2 Prototypes whose performance point has been achieved with working fluid having isentropic index $\gamma = 1.66$. The only valid scaling is to an He-charged derivative

		GPU-3 (He)	400 hp/cyl	V-160	PD-46
Parameter values of prototype gas path which transfer directly to derivative:					
N_T		3.276	3.09	2.727	2.67
γ		1.66	1.66	1.66	1.66
κ_{eq}		~1.0	–	0.967	~1.0
α_{eq} (deg)		120	–	90	120
$\P_v$		0.75	0.582	0.685	0.69

Values of dead space ratios δ_{xe}, δ_r, δ_{xc} transfer by default via scaled products $(\tfrac{1}{4}\pi)d_x^2 L_x n_{Tx}/V_{sw}$ (see below).

Raw prototype data having dimensions:					
Pwr	W	4250	291 000	8000	3000
rpm	1/min	2000	452	1500	3000
V_{sw}	cc	118.63	17 400	225	77.5
p_{ref}	bar	69	110	150	102
L_{xe}	mm	245.3	1090	240	104
L_r	mm	22.6	75	30	20.32
L_{xc}	mm	46.1	314	100.0	66
d_{xe}	mm	3.02	10.0	2.7	1.83
r_{hr}	mm	0.0304	0.0487	0.0279	0.02217
d_{xc}	mm	1.08	3.0	1.25	1.016
A_{ffr}	mm²	2406	69 258	2289	2796
n_{Txe}	–	40	49	24	96
n_{Txc}	–	312	750	302	152
R	J/kgK	2080	2080	2080	2080
μ_0	Pas	0.017E–03	0.017E–03	0.017E–03	0.017E–03

Lengths of regenerator and compression exchanger scale by the same ratio:

$$L_r^* = 22.6 \times 0.6334 = \mathbf{14.31\ mm}$$

$$L_{xc}^* = 46.1 \times 0.6334 = \mathbf{29.19\ mm}$$

The proportionality embodied in *FastTrack* applies in *FlexiScale*: $r_{hx}^*/L_x^* = r_{hx}/L_x$. Equally $d_x^*/L_x^* = d_x/L_x$. The same multiplier (0.6334) therefore scales the internal diameters:

$$d_{xe}^* = 3.02 \times 0.6334 = \mathbf{1.91\ mm}$$

$$d_{xc}^* = 1.08 \times 0.6334 = \mathbf{0.684\ mm}$$

$$r_{hr}^* = 0.0304 \times 0.6334 = \mathbf{0.01926\ mm}$$

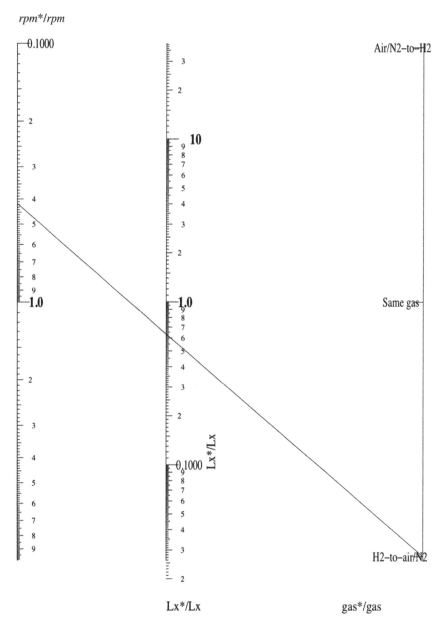

Figure 13.1 *Noting that scale values increase downwards* locate value of ratio *rpm*/rpm* on left-hand scale. Join with straight line to gas*/gas scale at relevant point – that is, H$_2$-to-air. Read exchanger length ratio L_x*/L_x from centre scale – in this specific example (only) *0.6334*. Determine length L_x* of derivative exchanger by multiplying up: L_x* = (L_x*/L_x)L_x

Numerical values to this point are in accurate agreement with those previously acquired using *FastTrack* – with the apparent exception of L_{xc}^* (and d_{xc}^*): Reports on the GPU-3 cite two values for the L_{xc}^*: a length wetted externally by the coolant of 35.5 mm, and an overall tube length of 46.1 mm. Previous numerical work by *mRT* – including the scaling exercise in *FastTrack* – have been based on the former of the two values. On the other hand, it is beyond doubt that the entire 46.1 mm length is involved in flow friction. Moreover, the header plates were wetted by coolant and brazed or soldered to the tubes, providing an effective thermal conduction path. Thus it is unlikely that the tube entry and exit sections were at a temperature greatly different from the nominal. On this basis, scaling proceeds using $L_{xc}^* = 29.19$ mm.

13.3   Operating conditions

To establish charge pressure p_{ref}^* of the derivative design, Figure 13.2 is entered via the left-hand scale at the value of the speed ratio already calculated, viz, at 0.4167 [–]. Noting the changed location of the target (H_2 to air) on the right-hand scale, join this point with a straight line to $rpm^*/rpm = 0.4167$ and from the centre scale read $p_{ref}^*/p_{ref} = 0.885$ [–]. Prototype p_{ref} being 69 bar (Table 13.1), derivative charge pressure is given by:

$$p_{ref}^* = 0.885 \times 69 \text{ bar} = \textbf{61.09 bar}$$

At the stage of setting derivative swept volume V_{sw}^* *FlexiScale* offers an insight – trivial once perceived, but fundamental to the whole business of scaling and to Stirling engine design in general. First the arithmetic.

Using numerical values now known, form the compound ratio $(p_{ref}^*/p_{ref})(rpm^*/rpm) = 0.885 \times 0.4167 = 0.3688$ [–]. The point is located on the right-hand scale of Figure 13.3. Any straight line through this point intersects the other two scales in values of pwr^*/pwr and V_{sw}^*/V_{sw} achieving similarity of prototype and derivative gas process cycles.

According to Figure 13.3 there is scope for, say, quadrupling target power by quadrupling swept volume V_{sw} with no change to lengths L_x or hydraulic radii r_{hx}. Should this come as a surprise it is necessary only to recall that the *number* of exchanger tubes n_{Tx} and regenerator free-flow area A_{ffr} have still to be set: as well as four-times V_{sw}, the power increase is going to call for four-times n_{Tx} and four-times A_{ffr}. If still not convinced, visualize the factor of four being achieved in four identical gas path units – that is, in a four-cylinder engine.

The earlier exercise in *FastTrack* set $pwr^* = 1$ kW. The equivalent in *FlexiScale* calls for the line on Figure 13.3 to pass through point 1 kW/8.95 kW = 0.1117 [–] of the left-hand scale for pwr^*/pwr, and thus through the value 0.303 [–] on the centre scale for V_{sw}^*/V_{sw}. Multiplying up:

$$V_{sw}^* = 0.303 \times 118.63 \text{ cc} = \textbf{36.0 cc}$$

Design of the tubular exchangers is completed by determining respective numbers n_{Tx} of tubes. Expansion and compression exchangers are dealt with simultaneously: The numerical value $L_x^*/L_x = 0.6334$ [–] is the same for both, and is located on the right-hand scale of Figure 13.4. A straight line through this and the value for V_{sw}^*/V_{sw} (that is, 0.303) establishes a point on the un-graduated turning axis. The current value of d_x^*/d_x is identical to that for L_x^*/L_x, and

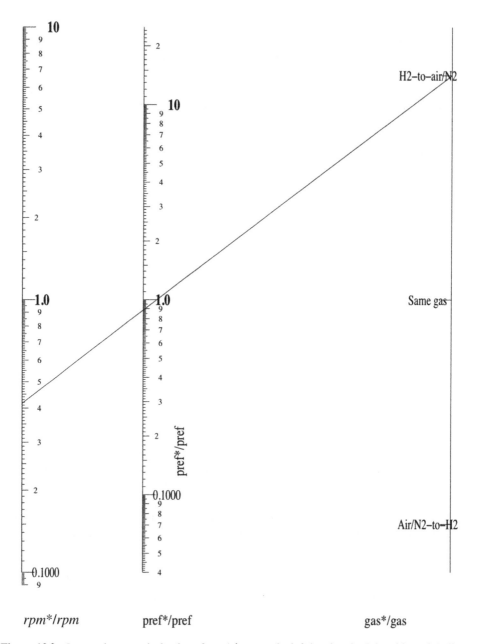

rpm*/rpm   pref*/pref   gas*/gas

Figure 13.2 Locate the numerical value of *rpm**/*rpm* on the left-hand scale. Join with straight line to right-hand scale at point corresponding to proposed working fluid change – in this case, H$_2$ to air. Read ratio p_{ref}*/p_{ref} from centre scale

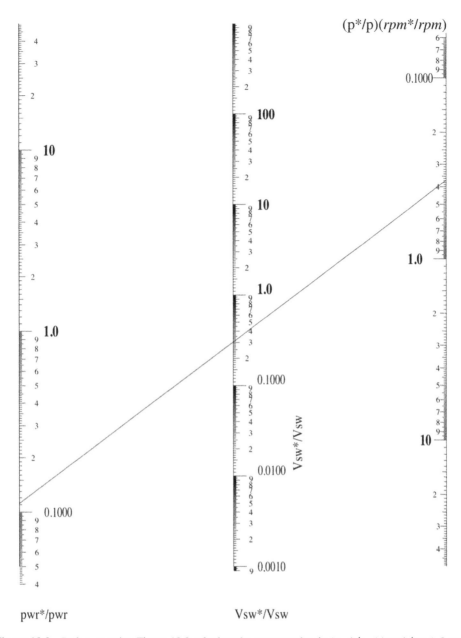

pwr*/pwr   Vsw*/Vsw

Figure 13.3   Before entering Figure 13.3 calculate the compound ratio $(p_{ref}*/p_{ref})(rpm*/rpm)$. Locate this value on right-hand scale *noting that scale values increase downwards*. On left-hand scale locate power ratio $Pwr*/Pwr$ and join with straight line. From centre scale read swept volume ratio $V_{sw}*/V_{sw}$

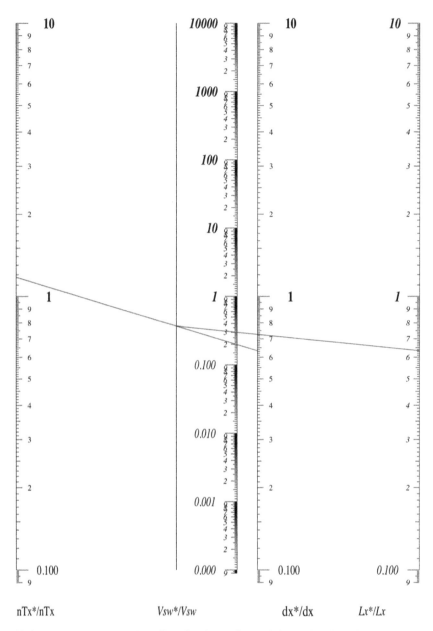

nTx*/nTx   Vsw*/Vsw   dx*/dx   Lx*/Lx

Figure 13.4 Connect (known) value of length ratio L_x^*/L_x on right-hand scale to (known) swept volume ratio on V_{sw}^*/V_{sw} scale, marking point of intersection of extended line with un-graduated turning scale. Connect this latter point to (known) diameter ratio d_x^*/d_x and extrapolate to scale for n_{Tx}^*/n_{Tx} and read off required value

is located on the d_x^*/d_x scale. A line from this latter point through the point on the turning axis intersects the n_{Tx}^*/n_{Tx} scale at the value 1.175 [–]. Multiplying up:

$$n_{Txe}^* = 1.175 \times 40 = \textbf{47 tubes}$$
$$n_{Txc}^* = 1.175 \times 312 = \textbf{367 tubes}$$

13.4   Regenerator matrix

Free-flow area A_{ffr} of the regenerator is acquired by equating dead volume ratios δ_r^* of derivative to that of prototype. By definition $\delta_r = A_{ffr}L_r/V_{sw}$ – in other words, $A_{ffr}^*L_r^*/V_{sw}^* = A_{ffr}L_r/V_{sw}$, or $A_{ffr}^* = A_{ffr}(V_{sw}^*/V_{sw})(L_r^*/L_r)^{-1}$. Using values already acquired for the ratio terms, and taking the value of A_{ffr} from Table 13.1:

$$A_{ffr}^* = A_{ffr}(V_{sw}^*/V_{sw})(L_r^*/L_r)^{-1} = 2406 \,[\text{mm}^2] \times 0.303 \times 0.6334^{-1} = \textbf{1150 mm}^2$$

Derivative housing cross-sectional area A_{ffH}^* is free-flow area divided by volume porosity $\P_v[-]$, this latter being held to the value 0.75 [–] by the GPU-3 specification.

$$A_{ffH}^* = A_{ffr}^*/\P_v = 1150\,[\text{mm}^2]/0.75 = \textbf{1533 mm}^2$$

It remains to convert hydraulic radius r_{hr} to wire diameter d_w [mm] mesh number m_w [wires/mm or wires/inch]. Invoking the standard definition for d_w in terms of r_{hr} and $\P_v$:

$$d_w = 4r_{hr}[\text{mm}](1 - \P_v)/\P_v \,[\text{mm}] = \textbf{0.0253 mm, or 0.001 inch}$$

$$m_w = \P_v/\pi r_{hr}[\text{wire/mm}] = \textbf{12.49 wire/mm or 317 wire/inch} \,(\approx \textbf{320 mesh})$$

To achieve dynamic similarity of the flow processes the regenerator must be stacked with square-weave wire gauze of the same volume porosity as that of the GPU-3. G Bopp and Co. Ltd. remains the recommended supplier of precision wire gauzes.

13.5   Rationale behind *FlexiScale*

As start point *FlexiScale* takes conditions in the wire-gauze regenerator as implied in the conventional steady-flow heat transfer and flow friction correlations – $StPr^{2/3}$ versus Re and Cf versus Re. By contrast with the respective pipe-flow counterparts, these are not mutually parallel. Reynolds analogy does not apply, and similarity of fractional pressure drop is not synonymous with similarity of fractional temperature profile, $\Delta T(x)/\Delta T_{in/out}$.

A close fit to the experimental $StPr^{2/3}$ versus Re correlation curve is:

$$StPr^{2/3} \approx a/Re^b \tag{13.1}$$

For Cf versus Re:

$$Cf \approx c/Re + d \tag{13.2}$$

Scaling exploits the correlations through the definitions of NTU and $\Delta p/p$ – both already derived:

$$NTU \approx StL_x/r_h$$
$$\approx aRe^{-b}L_x/r_h \tag{13.3}$$

$$\Delta p/p \approx (-)\tfrac{1}{2}\gamma Ma^2 CfL_x/r_h$$
$$\approx (-)\tfrac{1}{2}\gamma Ma^2 (c/Re + d)L_x/r_h \tag{13.4}$$

Equation 13.3 reflects the common approximation $Pr^{\frac{2}{3}} \approx$ unity.

The condition for both NTU and $\Delta p/p$ to be similar between prototype and derivative is that L_x/r_h, Ma, and Re be independently similar. In terms of the now familiar similarity parameters and similarity variables:

$$Ma = \frac{4\sigma'_\varphi}{\psi_\varphi} \frac{N_{MA}(L_x/V_{sw}^{\frac{1}{3}})}{\delta_x} \sqrt{(\tau/\gamma)}$$

$$= \frac{4\sigma'_\varphi}{\psi_\varphi} \frac{N_{Mx}}{\delta_x} \sqrt{(\tau/\gamma)} \tag{13.5}$$

$$N_{Mx} = \omega L_x/\sqrt{(RT_C)}$$

The similarity condition is $N_{Mx}{}^* = N_{Mx}$, where the asterisk denotes the derivative term. Prototype and derivative values of specific mass rate σ'_φ and of specific pressure ψ_φ are common for all crank angle φ. τ is T/T_C, where T is nominal exchanger temperature. τ^* cancels with τ on scaling between like exchangers. Expanding the expression for N_{Mx}, using the fact that $\omega = 2\pi rpm/60$ and cancelling terms whose respective numerical values are common between prototype and derivative:

$$rpm^* L_x{}^*/R^* = rpm L_x/R \tag{13.6}$$

Figure 13.1 is the graphical embodiment of Equation 13.6.

In terms of similarity parameters and variables, with account of the definition of N_{Mx}, the local instantaneous Reynolds number Re may be written:

$$Re = \frac{4\sigma'}{L_x/r_h} \frac{N_{Mx}^2}{\delta_x} \frac{N_{SG}}{f\{\tau(\lambda)\}}$$

The term $f\{\tau(\lambda)\}$ serves as a temperature coefficient for dynamic viscosity, μ. If an innocuous approximation is acceptable, it will disappear by cancellation during scaling.

Similarity of N_{Mx} has already been arranged, so $N_{Mx}{}^*$ will cancel N_{Mx} on eventually equating Re^* to Re. Fractional dead space δ_x belongs among the most basic similarity parameters, and will likewise cancel. Equating remaining terms, $N_{SG}{}^*/f\{\tau(\lambda)\}^* = N_{SG}/f\{\tau(\lambda)\}$.

The appropriate reference value of μ, viz. the μ_0 in each case, is already embodied in the definition of $N_{SG} = p_{ref}/\omega\mu_0$. While respective μ_0 differ by a factor of two between the three gases of interest, the temperature coefficients are closely similar: μ at 600 °C for air, for example, is a factor of 2.29 greater than the ambient value. That for He increases in the same ratio. The for H_2 the ratio between the same temperature is 2.17 [–]. On this basis temperature coefficients $f\{\tau(\lambda)\}$ cancel, leaving $N_{SG}^* = N_{SG}$. Substituting $\omega = 2\pi rpm/60$:

$$p_{ref}^*/rpm^* \mu_0^* = p_{ref}/rpm\mu_0 \tag{13.7}$$

Figure 13.2 is the graphical equivalent of Equation 13.7.

When considering the tubular exchanger, $L_x/4r_h$ gives way to L_x/d_x, since for the cylindrical duct $r_h = \frac{1}{4}d$. When applied to the regenerator, the numerical value of L_r/r_{hr} is used. Confusion does not arise, since expansion exchanger scales to expansion exchanger, compression to compression, and regenerator to regenerator.

14

ReScale

14.1 Introduction

FastTrack and *FlexiScale* both afford 'rigorous' scaling – rigorous in the sense that respective cycle histories of local *Re* and *Ma* of the prototype are faithfully replicated in the derivative – at least as seen in terms of the concepts of steady, 'one-dimensional' or 'slab' flow. When scaling to air/N_2 at modest *rpm* from a high-*rpm* prototype charged with H_2 at high-pressure, rigour comes at a cost: charge pressure p_{ref} is unattractively high. This is at odds with the perceived niche for the air/N_2-charged type, where p_{ref} between 5 and 15 bar seems appropriate.

If derivative values of flow-passage length $L_{xe}*$ and $L_{xc}*$ are acceptable, then there is a way forward. Neither *FastTrack* nor *FlexiScale* dwelt directly on the matter, but the derivative from either process is self-evidently a *virtual prototype*: As such it can serve for a further scaling stage – re-scaling – to p_{ref} of choice. The original scaling stage targeted air/N_2, which by any logic remains the derivative working fluid. The fact relieves the re-scale algebra of working between different pairs of values of R and μ.

The algebra of *ReScale* is sufficiently sparse to be a practical proposition for the hand calculator. However, there is a subtlety: while the exchangers scale unambiguously (respective *NTU* and $\Delta p/p$ of prototype achieved in derivative) the designer must choose between replicating *either NTU or* $\Delta p/p$ in the regenerator.

Both options are illustrated below – *NTU* matching and $\Delta p/p$ matching. As previously, the numerical example precedes exposition of the rationale.

14.2 Worked example step-by-step

The specification to be re-scaled – the 'virtual prototype' – will be the derivative of the scaling exercise carried out in Chapter 13 – *FlexiScale*. Target p_{ref} becomes 15 bar (abs), but the earlier *rpm* value (1500) is retained. 'Derivative' now signifies the specification to be re-scaled from the 'virtual' prototype. An asterisk (*) identifies re-scaled variable. Variables of the 'virtual' prototype (formerly the derivative) lose the asterisk they carried during *FastTrack* and *FlexiScale*.

Stirling Cycle Engines: Inner Workings and Design, First Edition. Allan J Organ.
© 2014 John Wiley & Sons, Ltd. Published 2014 by John Wiley & Sons, Ltd.

If the re-scale sequence works out, the result will be a specification embodying potential for the Beale modulus value of virtual prototype. Taking into consideration that *rpm* are now common at 1500, $V_{sw}*[cc] = V_{sw}[cc](p_{ref}/p_{ref}*)$:

$$V_{sw}* = 36.0[cc](61.09/15.0) = \textbf{146.6 cc}$$

If the hardware design stage is reached $V_{sw}*$ might be rounded to 150 cc. (In the internal combustion engine tradition, rounding might be to 149 cc.)

The new reference length $L_{ref}*$ is $V_{sw}*^{1/3}$:

$$L_{ref}* = (146.6E - 06)^{1/3}[m] = \textbf{0.0527 m}$$

The speed parameter $N_{MA}*$ is calculated for the derivative as $\omega V_{sw}*^{1/3}/\sqrt{RT_C}$, Stirling parameter as $N_{SG}* = p_{ref}*/\omega\mu$. The ω used is $2\pi \times rpm/60 = 157.08$ rad/sec:

$$N_{MA}* = 157.08 \times (146.6E - 06)^{1/3}/\sqrt{(287.0 \times 333.0)} = \textbf{0.0277 [--]}$$

$$N_{SG}* = 15.0E + 05/(157.08 \times 0.017E - 03) = \textbf{5.617E + 08 [--]}$$

14.2.1 Tubular exchangers

Internal diameter $d_{xe}*$ of the individual tube of the expansion exchanger is calculated from $d_{xe}* = d_{xe}(p_{ref}*/p_{ref})^{-0.166}$. Exponent 0.166 is the numerical value of $b/(b + 1)$, b being the index of the steady-flow heat transfer correlation $StPr^{2/3} = a/Re^b$. For the cylindrical duct and the turbulent regime $b = 0.2$:

$$d_{xe}* = 1.91 \text{ mm} \times (15/61.09)^{-0.166} = \textbf{2.4175 mm}$$

Compression tube internal diameter $d_{xc}*$ is calculated in the same way:

$$d_{xc}* = 0.684 \text{ mm} \times (15/61.09)^{-0.166} = \textbf{0.08645 mm}$$

Numbers of tubes n_{Tx} are calculated by inverting the equation for enclosed dead volume V_{dx}. In terms of dead space ratio $\delta_x = V_{dx}/V_{sw}$, expansion exchanger dead space ratio is $\delta_{xe} = \frac{1}{4}\pi n_{Txe}d_{xe}^2 L_{xe}/V_{sw}$

$$n_{Txe} = 4V_{sw}\delta_{xe}/(\pi d_{xe}^2 L_{xe}) = 0.592 \times 146.6 \times 10^{-6}/(\frac{1}{4}\pi \times 2.4175^2 \times 155.4 \times 10^{-9}) = \textbf{121}$$

Similarly, for the compression exchanger:

$$n_{Txc} = 0.118 \times 146.6 \times 10^{-6}/(\frac{1}{4}\pi \times 0.8645^2 \times 31.1 \times 10^{-9}) = \textbf{948}$$

The relatively large number of tubes suggests exploring the possibility of equivalent slots.

The steps completed so far ensure achievement in the re-scaled derivative of the cycle histories of *NTU* in the exchangers and regenerator of the original prototype (in this case, the GPU-3) and, evidently, of the 'virtual' prototype. For expansion and compression exchangers *only*, replication of respective fractional pressure drops $\Delta p/p$ is also achieved.

14.2.2   Regenerator

Dealing with the regenerator is going to call for a decision: The re-scaled derivative can inherit regenerator cycle temperature history (by retaining prototype NTU_r) or it can inherit the cycle history of $\Delta p/p$ – but in general it cannot inherit both. Taking the former option leaves the $\Delta p/p$ history to 'float'. Taking the latter leaves NTU floating. There is the intriguing possibility that either $\Delta p/p$ or NTU – or both – may shift to a more favourable range. Equally, either option or both may lead to the less promising alternative.

It is a feature of scaling that respective numerical magnitudes of NTU, $\Delta p/p$ (and, indeed, of Ma, Re) remain unknown: For successful replication of prototype conditions, the information appears to be needed at this point. Where it is to come from?

A moment's reflection confirms that the ostensible need is actually served by a value of a *ratio* – NTU^*/NTU or $(\Delta p/p)^*/(\Delta p/p)$. The scaling algebra readily delivers the required ratios. If NTU^* have floated to a range such that $NTU^*/NTU >$ unity while $(\Delta p/p)^*/(\Delta p/p)$ remains unchanged, it is a reasonable conclusion that the gas process cycle has benefited. For similar assurance, $(\Delta p/p)^*$ must float to a range which leaves $(\Delta p/p)^*/(\Delta p/p) <$ unity (with NTU^* unchanged).

14.2.2.1   Scaling to preserve prototype $\Delta p/p$

For derivations, see Section 14.4 below – *Rationale behind* ReScale.

$$L_r^* = L_r(p_{ref}/p_{ref}^*)^{\frac{1}{4}} = 14.31 \text{ [mm]} \times (61.09/15.0)^{\frac{1}{4}} = \mathbf{20.33 \text{ mm}}$$

$$r_{hr}^* = r_{hr}(L_r^*/L_r)^3 = 0.01926 \text{ [mm]} \times (20.33/14.31)^3 = \mathbf{0.05523 \text{ mm}}$$

These values cause the derivative to inherit the $\Delta p/p$ history of the prototype – but not, in general, the NTU history. The earlier scaling step retained the Re_r history of the prototype (see Section 8.5), leading to a simple expression for the NTU ratio:

$$NTU^*/NTU = (L_r^*/L_r)(r_{hr}/r_{hr}^*) = (20.33/14.31)(0.01926/0.05523) = \mathbf{0.495 \text{ [–]}}$$

Regenerator NTU have more than halved. But before this scaling option is dismissed it is worth noting that, above a threshold NTU, increasing regenerator NTU brings rapidly diminishing returns in terms of temperature recovery ratio. Were it known that the GPU-3 value is above that threshold, there might be no problem in accepting the reduction now contemplated.

It is not necessary to grapple with the decision at this stage, since the implications of the alternative – of retaining the GPU-3 value of NTU – have not yet been explored.

14.2.2.2   Scaling to preserve prototype NTU_r

Hydraulic radius r_{hr}^* is calculated by the formula already used for d_{xe} and d_{xc} – but with index b appropriate to the volume porosity of the wire gauze used in the GPU-3. That value is $b = 0.385$ [–] and yields an exponent of -0.2779:

$$r_{hr}^* = 0.01926 \text{ [mm]}(15/61.09)^{-0.2779} = \mathbf{0.0286 \text{ mm}}$$

Retention of virtual prototype lengths L_x applied to the tubular exchanger *only*. For the regenerator $L_r{}^*$ has to be re-calculated using $L_r{}^* = L_r(r_{hr}{}^*/r_{hr})^{1/3}$:

$$L_r{}^* = L_r(r_{hr}{}^*/r_{hr})^{1/3} = 14.31[\text{mm}](0.0286/0.01926)^{1/3} = \mathbf{16.32\ mm}$$

The increase in $L_r{}^*$ over the L_r of the virtual prototype brings about a reduction in intensity of the temperature gradient. This might be considered advantageous.

The ratio of fractional pressure drops is:

$$(\Delta p/p)^*/(\Delta p/p) = 0.6734\frac{c/Re^* + d}{c/Re + d} \tag{14.1}$$

Neither Re^* nor Re is known. However, if the numerical value of their *ratio* is known, then Equation 14.1 may be explored for a range of arbitrary values of either Re^* or Re. Section 14.4.2 develops the expression for the Re ratio:

$$Re^*/Re = \frac{p_{ref}{}^*}{p_{ref}}\frac{r_{hr}{}^*}{r_{hr}} \tag{14.2}$$

$$= (15.0/61.09)(0.0286/0.01926) = \mathbf{0.365\ [-]}$$

Setting $Re = 1000$ gives $Re^* = 365$. Substituting into Equation 14.1:

$$(\Delta p/p)^*(\Delta p/p) = 0.6734\frac{64/365 + 0.25}{64/1000 + 0.25} = \mathbf{0.912\ [-]}$$

With this choice of Re pumping loss has fallen to some 90% of prototype value, while loss due to heat transfer deficit remains unchanged. According to Equation 14.1, the break-even value of Re^* – the value at which $\Delta p/p^* = \Delta p/p$ – is 241 [–].

The reality is that Reynolds number swings through a wide range over a cycle. For very small values Equation 14.1 reduces to:

$$(\Delta p/p)^*/(\Delta p/p) \approx 0.6734\ Re/Re^* \tag{14.1a}$$

In that extreme case:

$$(\Delta p/p)^*/(\Delta p/p) \approx 0.6734/0.365 = \mathbf{1.845\ [-]}$$

Instantaneous fractional pumping power in the derivative now exceeds that of the prototype.

All other things being equal, low Re occurs at low volume rate Q', where instantaneous pumping power, $(\Delta pQ'\ [\text{W}])$ is low. As Re increases indefinitely, pressure drop ratio according to Equation 14.1 tends to the value 0.6734 [–], and fractional pumping power in the derivative at any instant is 67% of that of the prototype.

It is a matter of personal persuasion as to whether design proceeds in terms of the $L_r{}^*$ and $r_{hr}{}^*$ which preserve prototype regenerator $\Delta p/p$, or in terms of those which preserve NTU_r and thus the temperature solution. A relevant consideration is that the break-even value $Re^* = 241$

is a low figure which will be exceeded over much of the cycle. On this reasoning the pumping penalty of the air-charged derivative is a lower fraction of net output than for the prototype. The specimen arithmetic accordingly proceeds on the basis of the L_r^* and r_{hr}^* which preserve prototype NTU_r.

It remains to convert these values to a form which specifies a regenerator matrix.

14.3   Regenerator matrix

Free-flow area A_{ffr} is acquired by equating dead volume ratios δ_r^* of derivative to the (known) prototype value. Either the original prototype or the intermediate virtual prototype will serve. By definition $\delta_r = A_{ffr}L_r/V_{sw}$ – in other words, $A_{ffr}^*L_r^*/V_{sw}^* = A_{ffr}L_r/V_{sw}$, or $A_{ffr}^* = A_{ffr}(V_{sw}^*/V_{sw})(L_r^*/L_r)^{-1}$. Taking the GPU-3 value of A_{ffr} from Table 8.1 of Chapter 8 *FlexiScale*:

$$A_{ffr}^* = A_{ffr}(V_{sw}^*/V_{sw})(L_r^*/L_r)^{-1} = 2406[\text{mm}^2] \times (146.6/118.63) \times (22.6/14.3)^{-1} = \mathbf{1881\ mm^2}$$

Derivative housing cross-sectional area A_{ffH}^* is free-flow area divided by volume porosity $\P_v[-]$, this latter being held to the GPU-3 value of 0.75 [–].

$$A_{ffH}^* = A_{ffr}^*/\P_v = 1881[\text{mm}^2]/0.75 = \mathbf{2508\ mm^2}$$

It remains to convert hydraulic radius r_{hr} to wire diameter d_w [mm] mesh number m_w [wires/mm or wires/inch]. Invoking the standard definition for d_w in terms of r_{hr} and $\P_v$:

$$d_w = 4r_{hr}\ [\text{mm}]\ (1 - \P_v)/\P_v\ [\text{mm}] = \mathbf{0.03813\ mm},\ \text{or}\ \mathbf{0.0015\ inch}$$

$$m_w = \P_v/\pi r_{hr}\ [\text{wire/mm}] = \mathbf{8.347\ wire/mm}\ \text{or}\ \mathbf{212\ wire/inch}\ (\approx \mathbf{210\ mesh})$$

14.4   Rationale behind *ReScale*

Both *FastTrack* and *FlexiScale* achieve replication of the gas processes through the 'obvious' expedient of equating derivative Re and Ma and the three L/r_h to respective values of these quantities in the prototype. Doubts as to the relevance – or otherwise – of steady-flow correlations did not arise. Parallel-duct exchangers and matrix regenerator are handled by a common arithmetic.

ReScale instead equates variations in exchanger $NTU_x(= StL_x/r_{hx})$ and $\Delta p/p$ ($\propto N_{Mx}^2 NTU_x$).

14.4.1   Tubular exchangers

It is assumed (a) that the correlations for incompressible, steady-flow are relevant, and (b) that flow is turbulent, allowing $St = a/(Pr^{2/3}Re^b) = \tfrac{1}{2}Cf$. Ensuring similarity of both NTU_x and $\Delta p/p$ starts by holding prototype and derivative N_{Mx} the same. N_{Mx} is defined $\omega L_x/\sqrt{(RT_C)}$. The gas is unchanged (R the same), as are *rpm* (and thus ω), so the required condition is met by retaining the values of L_{xe} and L_{xc} which resulted from initial scaling using *FastTrack* or *FlexiScale*.

Achieving commonality of NTU_x (and thus of $\Delta p/p$) is now a case of equating:

$$a/Re^b(L_x/d_x) = a\left\{\frac{4\sigma'}{L_x/d_x}\frac{N_{Mx^2}}{\delta_x}\frac{N_{SG}}{f\{\tau(\lambda)\}}\right\}^{-b}L_x/d_x$$

The commonality of σ', of δ_x and of $f\{\tau(\lambda)\}$ is inherent. With N_{Mx} unchanged from the proto-type value, the similarity condition reduces to $(N_{SG}*)^{-b}/(L_x*/d_x*)^{1+b} = N_{SG}^{-b}/(L_x/d_x)^{1+b}$. Angular speed ($\omega$ and or *rpm*), coefficient of dynamic viscosity μ_0 and L_x are unchanged, leaving:

$$(p_{ref}*)^{-b}(d_x*)^{1+b} = p_{ref}^{-b}d_x^{1+b} \tag{14.3}$$

Equation 14.3 with b set equal to 0.2 is the similarity condition for both NTU_x and $\Delta p/p$ of the tubular exchangers.

14.4.2 Regenerator

The designer aims for similarity of NTU_r, or similarity of $\Delta p/p$. The former is achieved as for the tubular exchangers, but with the numerical value of coefficient b set to 0.385 [−]:

The condition for the alternative – similarity of $\Delta p/p$ – is seen by inspecting the defining equation for steady flow. For the wire screen matrix, friction factor Cf is a function of Re of the form $Cf = c/Re + d$:

$$\Delta p/p \approx (-)\tfrac{1}{2}\gamma Ma^2(c/Re + d)L_r/r_{hr} \tag{14.4}$$

Overall similarity is achieved if similarity of the compound term Ma^2L_r/r_{hr} is achieved independently of Re. Ma can be expressed in terms of dimensionless operating parameters and Finkelstein's dimensionless variable σ'_φ:

$$Ma = \frac{4\sigma'_\varphi}{\psi_\varphi}\frac{N_{Mr}}{\delta_r}\sqrt{(\tau/\gamma)} \tag{14.5}$$

This part of the scaling process may be thought of as setting to unity the ratio of derivative and prototype Ma^2L_r/r_{hr}. Algebraic simplification results on doing so explicitly, whereupon several terms of Equation 14.5 cancel. N_{Mr} is defined $\omega L_r/\sqrt{(RT_C)}$, so with ω, R, and T_C unchanged, similarity of Ma^2L_r/r_{hr} is achieved by similarity of $N_{Mr}^2L_r/r_{hr}$ – that is, by similarity of of L_r^3/r_{hr}:

$$r_{hr}*/r_{hr} = (L_r*/L_r r_{hr})^3 \tag{14.6}$$

Re-writing Equation 14.4 as a ratio of derivative to prototype terms:

$$(\Delta p/p)^*/(\Delta p/p) = \frac{(L_r^3/r_{hr})^*}{L_r^3/r_{hr}} \frac{c/Re^* + d}{c/Re + d}$$ (14.7)

Supposing a value were available for Re^*/Re, then for any arbitrary value of Re^*, Re would be known. It would then be possible to explore Equation 14.7 over a sufficient range of Re to be confident of the value of $(\Delta p/p)^*/(\Delta p/p)$.

$$Re = a \left\{ \frac{4\sigma'}{L_r/r_{hr}} \frac{N_{Mr}^2}{\delta_r} \frac{N_{SG}}{f[\tau(\lambda)]} \right\}$$

For $N_{Mr}^2/(L_r/r_{hr})$ may be substituted $\{N_{Mr}^2(L_r/r_{hr})\}/(L_r/r_{hr})$:

$$Re = a \left\{ \frac{4\sigma'}{(L_r/r_{hr})^2} \frac{N_{Mr}^2(L_r/r_{hr})}{\delta_r} \frac{N_{SG}}{f[\tau(\lambda)]} \right\}$$

Similarity of the term $N_{Mr}^2(L_r/r_{hr})$ has already been arranged. Upon cancelling terms which are common between virtual prototype and derivative (including components of $N_{SG} = p_{ref}/\omega\mu_0$ which are common) and recalling Equation 14.6, the ratio Re^*/Re is:

$$Re^*/Re = \frac{P_{ref}^*}{P_{ref}} \frac{(L_r/r_{hr})^2}{(L_r^*/r_{hr}^*)^2}$$ (14.8)

15

Less steam, more traction – Stirling engine design without the hot air

15.1 Optimum heat exchanger

An 'optimum' heat exchanger can be expected to have a lower exit temperature difference and lower pumping loss than an exchanger of arbitrary design under the same thermal load and inlet conditions. For the steady flow case the specification (flow passage length L_x, hydraulic radius r_h and number of flow channels in parallel n_{Tx}) yielding optimum performance for given operating conditions is readily identified. Application to the operating conditions of the Stirling engine remains a matter of judgement, and is subject to reservations, including the following:

> Ideally the thermal exchange process would be modelled in entirety – from combustion products via external and internal wall surfaces through to core temperature of the fluid flowing internally.
>
> To avoid embarking on a PhD programme it will be assumed that the internal surface temperature of the exchanger duct is is uniform and constant at T_w. If, as is possible (see Chapter 11, Section 11.4), the thermal bottleneck arises between the combustion gases and the *outside* surface of the exchanger, this assumption becomes questionable.
>
> Mechanized optimization is a numerical ritual which mindlessly exploits both strengths *and weaknesses* of the object function[1] in its search for parameter values which maximize it. If the object function is flawed, the resulting 'optimum' can be worse than an arbitrary design.

The operating conditions envisaged as optima here are those under which the total of *available work lost* to the net sum of (a) temperature deficit and (b) pumping is a minimum.

Temperature deficit is a function of *NTU* (Number of Transfer Units) $= 4StL_x/d_x$. Pumping loss is a function of friction factor – specifically of C_fL_x/d_x. For steady, fully-developed pipe flow in the turbulent regime $C_f \approx 2St$ – a fact which permits **both loss components to be expressed in terms of *NTU*.** Differentiating with respect to *NTU* the expression for net available

---

[1] Object function: a symbolic expression defining physical or abstract entity to be maximized (or minimized).

Stirling Cycle Engines: Inner Workings and Design, First Edition. Allan J Organ.
© 2014 John Wiley & Sons, Ltd. Published 2014 by John Wiley & Sons, Ltd.

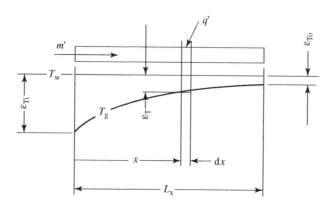

Figure 15.1   Temperature distribution in steady flow at uniform *NTU*

work lost and equating to zero yields an algebraic expression in NTU_{opt} having explicit solution. The length, internal diameter and number of tubes which achieve the numerical value of NTU_{opt} are available in algebraic combination which can be displayed as a design chart.

15.2   Algebraic development

Figure 15.1 represents one of a total of n_{Tx} ducts in parallel, each of length L_x [m], of internal diameter d_x [m] and at constant, uniform temperature T_w. Gas enters from the left at $x = 0$ with temperature T_i. Computation proceeds in terms of temperature *difference, or 'error'*, ε_T, defined as $T - T_w$, of which the value at inlet, ε_{Ti}, is $T_i - T_w$.

The standard solution for the variation of ε_T with fractional distance x/L_x is:

$$\varepsilon_T(x/Lx) = \varepsilon_{Ti} e^{-NTUx/Lx} \quad [K] \tag{15.1}$$

NTU in Equation 15.1 has already been defined. $St = $ Stanton number $h/\rho|u|c_p$.
At exit, $x/L_x = $ unity, and:

$$\varepsilon_{To} = \varepsilon_{Ti} e^{-NTU} \quad [K] \tag{15.1a}$$

A steadily flowing stream of ideal gas exchanges heat at rate $m'c_p(\varepsilon_{Ti} - \varepsilon_{To})$ [W]. This must equal mean rate $\underline{Q}'_x = dQ_x/dt$ [W] at which heat is exchanged externally to maintain wall temperature constant at T_w. Using Equation 15.1a to eliminate ε_{To} yields $\underline{Q}'_x = m'c_p\varepsilon_{Ti}(1 - e^{-NTU})$ [W]. Inverting to give inlet temperature difference ε_{Ti} in terms of $\underline{Q}'_x$:

$$\varepsilon_{Ti} = \underline{Q}'_x(1 - e^{-NTU})^{-1}/m'c_p \ [K] \tag{15.2}$$

$\underline{Q}'_x$ may be expressed in terms of shaft power *Pwr* and brake thermal efficiency η_{bth} [–]. For expansion and compression exchangers respectively:

$$\underline{Q}'_{xe} = Pwr/\eta_{bth} \quad [W] \tag{15.3a}$$

$$\underline{Q}'_{xc} = Pwr(1 - \eta_{bth})/\eta_{bth} \quad [W] \tag{15.3b}$$

The expansion exchanger provides the example: Equation 15.3a is substituted into Equation 15.2, giving:

$$\varepsilon_{Ti} = (Pwr/\eta_{bth})(1 - e^{-NTU})^{-1}/m'c_p \quad [\text{K}] \tag{15.4}$$

Temperature 'error' ε_T may now be expressed as a function of x/L_x [i.e., as $\varepsilon_T(x/L_x)$] on substituting Equation 15.4 into Equation 15.1.

Over the elemental length of duct dx (Figure 15.1), rate of loss of available work, $T_0ds'_{gen}$ may be evaluated (Bejan 1994), $q'T_0\varepsilon_T/T^2$:

$$T_0ds'_{gen} = q'T_0\varepsilon_T/T^2 \quad [\text{W}] \tag{15.5}$$

In Equation 15.5 q' [W] is heat rate by convection over channel length dx, T is local temperature and T_0 is ambient temperature – the lowest temperature at which q' could realistically be rejected. For present purposes T_0 can be replaced by T_C.

In terms of ε_T (now known), $q' = -h \cdot \pi d_x \varepsilon_T dx$, in which h [W/m^{2}K] is coefficient of convective heat transfer. Dividing the right-hand side by $\rho|u|c_p$ (and multiplying by $m'c_p/A_{ff}$ to compensate) exposes Stanton number, $St = h/\rho|u|c_p$:

$$q' = -\varepsilon_T m'c_p \, Stdx/r_h \tag{15.6}$$

By definition $Stdx/r_h = (StL_x/r_h)x/L_x = NTUx/L_x$. Making this substitution in Equation 15.6, substituting the result into Equation 15.5, noting that $c_p = R\gamma/(\gamma - 1)$ and integrating with respect to x over length L_x of the expansion exchanger ($T = T_E$):

$$T_0ds_{q'} = (+)\frac{\frac{1}{2}\gamma m'\varepsilon_{Ti}{}^2 RT_C}{(\gamma - 1)T_E{}^2}(1 - e^{-2NTU}) \tag{15.7}$$

Pressure drop $\Delta p = (-)\frac{1}{2}\rho u^2 C_f L_x/r_h$. Corresponding pumping power W'_f [W] $= \Delta p u A_{ff} = \frac{1}{2}\rho|u|^3 A_{ff} C_f L_x/r_h$ [W]. By Reynolds' analogy, $C_f L_x/r_h \approx 2StL_x/r_h$:

$$W'_f \approx \frac{1}{2}\rho|u|^3 2NTUA_{ff} \quad [\text{W}] \tag{15.8}$$

Availability Theory allows for conversion to useful work of friction heat generated at a temperature T above ambient ($T_0 = T_C$). For consistency with Equation 15.7:

$$T_0ds'_f \approx (T_0/T)\rho|u|^3 NTUA_{ff} \quad [\text{W}] \tag{15.9}$$

For expansion and compression exchangers the term T_0/T takes the respective values $1/N_T$ and unity.

Equation 15.4 is substituted into Equation 15.7, eliminating ε_{Ti}. The expressions for the two loss components are normalized by dividing by $p_{ref}V_{sw}$. (Beale modulus N_B is the result of normalizing cycle work W in the same way. The manoeuvre expresses losses and net cycle work to a common denominator (literally!), allowing direct comparison of useful and penalty values.

In the normalization process m' gives way to *specific mass rate* σ', which is a 'cycle invariant' (Chapter 10, Section 10.3). The resulting expression for net loss rate Z' is a function of NTU and of terms under the control of the designer. For the expansion exchanger:

$$Z'_{xe} = (T_0 ds_q' + T_0 ds_f')/p_{ref} V_{sw}$$
$$= \pi\sigma'^{-1} \underline{Q}_{xe'}^2 \{(\gamma - 1)/\gamma N_T^2\}(1 - e^{-NTU})/(1 + e^{-NTU})$$
$$+ \sigma'^3 NTU (N_T/\psi\delta_{xe})^2 (N_{MA} L_{xe}/L_{ref})^2 \qquad (15.10)$$

The value of NTU which minimizes Z'_{xe} is obtained by differentiating Equation 15.10 with respect to NTU and equating to zero. The result is a quadratic equation in e^{-NTU} having two roots of equal magnitude and opposite sign – which considerably facilities choice of the appropriate root!

For engines of interest N_T might typically be 3.3 [−]. Specific mass rate σ' and fractional pressure ψ are cycle invariants, and δ_{xe} for a viable design will not be far from 0.35. The parameters of the solution are $\underline{Q}'_{xe}$, N_{MA}, L_{xe}/L_{ref} and isentropic index γ. Combining N_{MA} and L_{xe}/L_{ref} into a compound dimensionless parameter $[N_{MA}(L_{xe}/L_{ref})]$ allows the complete solution for NTU_{opt} to be displayed in four charts – expansion exchanger, compression exchanger, each for both values of specific heat ratio $\gamma = 1.4$ and $\gamma = 1.66$. Two charts are featured in the section below, both for $\gamma = 1.4$, one for the expansion exchanger, one for the compression.

There is no optimum of L_x/L_{ref} in isolation, so the aim is now to establish the value of L_x/d_x – the 'aspect ratio' – which minimizes Z'_x by achieving NTU_{opt}. Proceeding in two stages has dramatically simplified the search for best L_x/d_x.

NTU – optimum or otherwise – is defined:

$$NTU = 4St L_x/d_x \qquad (15.11)$$

In turn:

$$St \approx aRe^b Pr^c \qquad (15.12)$$

Values used in computation are $a = 0.023$, $b = -0.2$, $c = -0.4$ [−].

Expressed in terms of the parameters of Stirling cycle analysis, the Reynolds number Re is:

$$Re = \frac{\sigma' N_{RE}(L_x/L_{ref})^2}{\delta_x(L_x/d_x)} \qquad (15.13)$$

Substituting Equation 15.13 into Equation 15.12 and the result into Equation 15.11 leads to an explicit expression for optimum L_x/d_x:

$$(L_x/d_x)^{1+b} = \tfrac{1}{4}a^{-1} NTU_{opt}(\sigma'/\delta_x)^b N_{RE}^b (L_x/L_{ref})^{2b} \qquad (15.14)$$

Equation 15.14 is of the form $f(L_x/d_x) = f(NTU_{opt}) f(N_{RE}) f(L_x/L_{ref})$. By taking logarithms this may be transformed to:

$$f_q(L_x/d_x) = f_u(NTU_{opt}) + f_v(N_{RE}) + f_w(L_x/L_{ref}) \qquad (15.15)$$

A function in this form can be converted to a nomogram, as used for display shortly.

15.3 Design sequence

The proposed design sequence applies to engine projects complying with outline specification of Table 15.1. It is a chart-based route to length, L_x, number, n_{Tx} and internal diameter d_x of exchanger tubes which is 'optimum' in the following sense: for operating conditions (p_{ref}, V_{sw}, rpm, working fluid) specified by the designer, the sum of available work lost per cycle to pumping and to heat transfer deficit is minimized. The minimum – and corresponding 'optimum' – reflect idealizations: in particular it worth keeping in mind that the entire approach is predicated on the presumption of relevance of steady-flow heat transfer and friction correlations ($StPr^{2/3} - Re$ and $C_f - Re$ correlations) to the unsteady conditions of the Stirling engine.

The approach is illustrated by designing expansion and compression exchanger tube-sets for an air-charged (or N_2-charged) engine targeting 1 kW.

Table 15.1 Arbitrary 'standard' specification for use in worked example. Dead-space ratios $\delta_r = V_{dr}/V_{sw}$ and so on lie within the range spanned by benchmark engines. An engine which fails to work with these values cannot be made to do so merely by changing the δ_x. Conversely, a plausible value of dead space can be exploited to accommodate a viable combination of hydraulic radius r_h, free-flow area A_{ff}, and flow-passage length L_x (or L_r)

Temperature ratio $N_T = T_E/T_C$: 3.3 (e.g., T_E = 1020 K, T_C = 310 K)
For opposed-piston or V-configuration:

thermodynamic (and kinematic) volume ratio κ: 1.0
thermodynamic (and kinematic) phase angle α: 90 degrees

Achieving these values in a coaxial, uniform-bore machine will call for the following values of kinematic phase angle β and piston/displacer displacement ratio λ:

β = 120 degrees
λ = 1.0

$\delta_e = V_{de}/V_{sw}$	0.10	[−]
$\delta_{xe} = V_{dxe}/V_{sw}$	0.35	[−]
$\delta_r = V_{dr}/V_{sw}$	0.40	[−]
$\delta_{xc} = V_{dxc}/V_{sw}$	0.15	[−]
$\delta_c = V_{dc}/V_{sw}$	0.15	[−]

volume porosity of regenerator matrix $\P_v$ 0.7 [−]

---

1. Select swept volume V_{sw} and realistic values for mean charge pressure p_{ref} and speed *rpm*, and apply Beale number criterion. With $V_{sw} = 175$ cc, $p_{ref} = 15$ bar (abs) and *rpm* = 1500:

$$N_B = \frac{1000 \ [\text{W}]}{15 \times 10^5 \ [\text{Pa}] \times 175 \times 10^{-6} \ [\text{m}^3] \times 1500/60 \ [\text{s}^{-1}]}$$

$$= 0.1524$$

The figure is well-known to be achievable. It is rounded down to 0.15, consistent with the likely precision of the method.

2. Set target brake thermal efficiency η_{th} – a modest 20% for this example. Calculate corresponding mean specific thermal loading Q'_x on the assumption that heat into and out of the cycle is exclusively via the tubular exchangers (working spaces adiabatic). For expansion and compression exchangers respectively:

$$Q'_{xe} = N_B/(2\pi\eta_{th}) = 0.15/(2\pi \times 0.2) = 0.119 \ [-]$$

$$Q'_{xc} = Q'_{xe}(1 - \eta_{th}) = 0.0955 \ [-]$$

3. Acquire a value for reference length $V_{ref} = V_{sw}^{1/3} = 175 \ [\text{cc}]^{1/3} = 5.593 \ [\text{cm}]$ or, in convenience units, 55.93 mm. Acquire values of dimensionless operating parameters. These can be read directly from the charts of Chapter 10: the speed parameter N_{MA} is read from Figure 10.3 – taking care to use the scale for air/N_2. The result is $N_{MA} = 0.0294$. From Figure 10.5 $N_{SG} = 0.562 \times 10^9$. From Figure 10.6 $N_{RE} = N_{SG}N_{MA}^2 = 4.87 \times 10^5$.

4. Select heated length of expansion exchanger duct L_{xe} starting from the numerical value of the ratio L_{xe}/L_{ref}. There is a reason for working backwards: the lower the value of L_{xe}/L_{ref} which can be converted into a practical heat exchanger (bearing in mind considerations of fabrication, external heat provision, etc.) the lower can be the sum total of losses due to pumping and heat transfer deficit. The reason should be clear from the analytical background. L_{xe}/L_{ref} value for the GPU-3 and USS P-40 is close to 5.0. This example will explore the possibility of achieving performance improvement by lowering L_{xe}/L_{ref} to 4.0. From the value of L_{ref} recently calculated, $L_{xe} = 4.0 \times 55.93$ mm $= 223.72$ mm or, say, 225 mm.

5. Calculate the product of L_{xe}/L_{ref} with $N_{MA} = 4.0 \times 0.0294 = 0.1176 \ [-]$, and locate this point on the with $N_{MA}L_{xe}/L_{ref}$ axis of Figure 15.2, which applies to gases for which isentropic index $\gamma = 1.4$ (e.g., air, N_2, H_2). The parameter curves are for values of the ratio Q'_{xe} in the range 0.5 to 5.0. The left hand scale gives the value of *NTU* (Number of Transfer Units, $4StL_{xe}/d$) for which net available work lost in the exchanger per cycle is a minimum *for the designer's proposed Q'_{xe} and L_{xe}/L_{ref}*.

   Values from the GPU-3 specification will provide context: $Q'_{xe} = 0.1311$; $N_{MA}L_{xe}/L_{ref} = 0.0817$. Working fluid is H_2 with $\gamma = 1.4$, so Figure 15.2 applies – with the reservation that the figure is based on the 'standard' specification (Table 15.1), which differs somewhat from that of the GPU-3. Interpolating within the $N_{MA}L_{xe}/L_{ref}$ and scale Q'_{xe} of Figure 15.2 suggests optimum *NTU* for the GPU-3 expansion exchanger would be about 2.8. Hargreaves (1991) uses symbol Λ for *NTU*. Converting his statement at his p. 433 to the present notation: '... *NTU should be larger than unity, preferably NTU > 2. Larger values are seldom practicable because the flow resistance and dead space are then excessive.*' NTU_{xe} back-calculated from the GPU-3 specification is 1.056, suggesting the exchanger design is less than optimum.

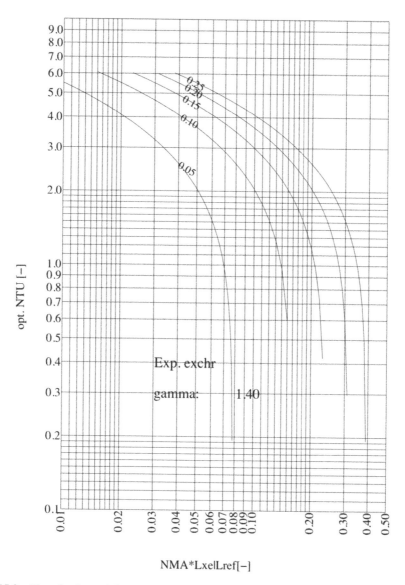

opt. NTU [-]

NMA*LxelLref[-]

Figure 15.2   Chart for determining optimum *NTU* in terms of product $N_{MA}L_{xe}/L_{ref}$ and dimensionless thermal loading Q'_{xe} (parameter curves) for expansion exchanger

Entering Figure 15.2 with values of $N_{MA}L_{xe}/L_{ref}$ and Q'_{xe} for the evolving design gives $NTU_{opt} \approx 2.0 - just~as~the~good~Dr.~Hargreaves~prescribed!$

6. Acquire length/diameter giving minimum net loss of available work: In Figure 15.3 use a straight line to join the NTU_{opt} value (2.0) of the left-hand scale to 4.87×10^5 on the N_{RE} scale. With another straight line join the chosen value (4.0) on the L_{xe}/L_{ref} scale to the point where the previous straight line intersected the (un-graduated) turning scale. Read the 'optimum' value of L_{xe}/d_{xe} at the point of intersection of the **second** straight line. The value is 132.0 With $L_{xe} = 225$ mm, d_{xe} follows as 225/132 = 1.704 mm.

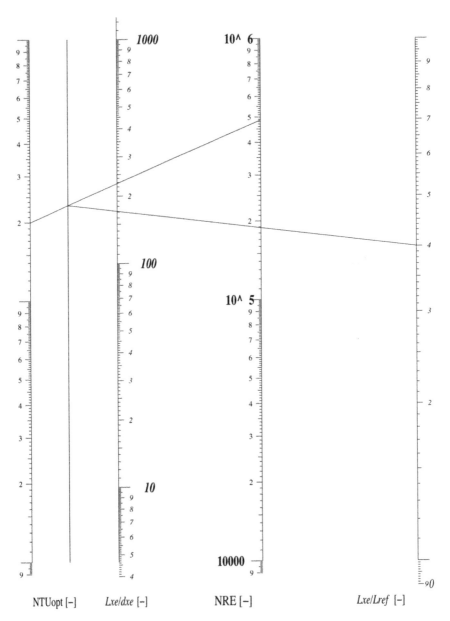

Figure 15.3 Optimum L_{xe}/d_{xe} in terms of N_{re} and L_{xe}/L_{ref} for expansion exchanger. *NTU* and N_{RE} scales can be connected, as can L_{xe}/d_{xe} and L_{xe}/L_{ref} (see text). No other connections allowed!

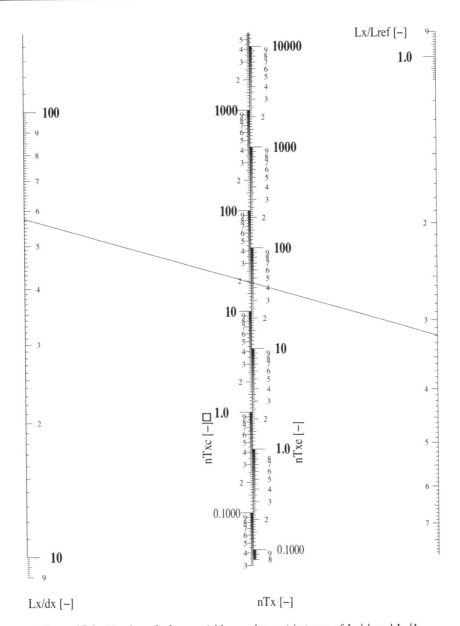

Figure 15.4   Number of tubes n_{Tx} (either exchanger) in terms of L_x/d_x and L_x/L_{ref}

7. Acquire number of tubes, n_{Txe}: In Figure 15.4 connect with a straight line $L_{xe}/d_{xe} = 132.0$ of the left-hand scale with $L_{xe}/L_{ref} = 4.0$ of the right-hand scale, and from the n_{Txe} side of the centre scale (right-hand side) read number of tubes $n_{Txe} = 120$.

In principle, design of the compression exchanger proceeds in the same way, but starting from Figure 15.5 with figures for Q'_{xc} (= 0.0955) and $N_{MA}L_{xc}/L_{ref}$. To provide context,

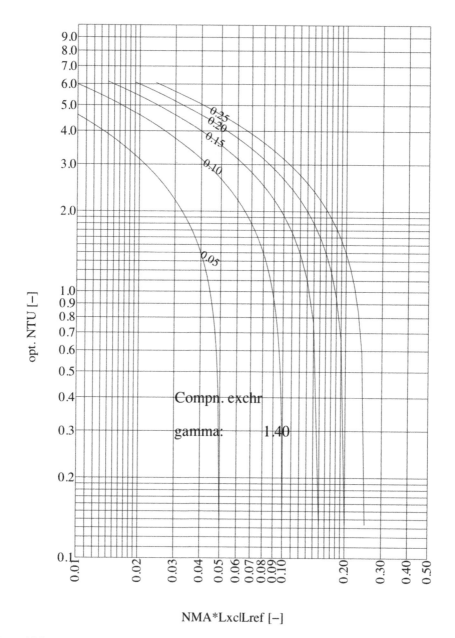

NMA*LxclLref [−]

Figure 15.5   Chart for determining optimum *NTU* in terms of product $N_{MA}L_{xc}/L_{ref}$ and dimensionless thermal loading, Q'_{xc} (parameter curves) for compression exchanger

GPU-3 values of L_{xc}/L_{ref} and NTU_{xc} are respectively 0.94 [–] and 0.525 [–]. '*In principle*' because internal gas path design must be reconciled with machining, fabrication and packaging considerations. The computational sequence is modified to handle the resulting compromise.

4b Select a value of L_{xc}/L_{ref} leading to L_{xc} which can be accommodated within proposed engine geometry. Taking $L_{xc}/L_{ref} = 2.0$ gives $L_{xc} = 2.0 \times 55.93$ mm $= 118$ mm or, say, 120 mm – accepted here because it fits a 1 kW concept under development by *mRT*.

5b The product of L_{xc}/L_{ref} with $N_{MA} = 2.0 \times 0.0294 = 0.0588$ [–]. Recalling that $Q'_{xc} = 0.0955$ [–] and interpolating in Figure 15.5 suggests that optimum NTU are 1.5 [–].

6b Using Figure 15.6 as previously explained for Figure 15.3 gives $L_{xc}/d_{xc} = 120$ [–]. With $L_{xc} = 120$ mm, $d_{xc} = 1.0$ mm.

So far so good!

7b Remembering to use the *left-hand* graduations of the centre scale, refer to Figure 15.4 to acquire $n_{Txc} = 380$ tubes.

The number is of the same order as n_{Txc} for the GPU-3, USS P-40, and USS V-160 (respectively 312, 400, and 302). This does not improve a proposition which is unattractive in terms of both fabrication and likely cost. The conversion chart at Figure 11.5 offers comparable NTU from the equivalent slotted exchanger: at four slots/tube there are 95 slots $b = 0.5$ mm wide, $h = 6.25$ mm deep ($h/b = 12.5$ [–]). The MP1002CA had 160 slots per exchanger of average aspect ratio $h/b = 4.95$ [–].

From a thermodynamic point of view, either exchanger – tubular or slotted – promises improvement over the corresponding component of the GPU-3, for which back-calculated NTU are 0.525 [–], or one-third of 'optimum'. (The algebra of the back-calculation is identical to that behind the design charts, so the comparison is reliable.)

Practical design is all about reconciling incompatibilities – about compromise. Accepting this suggests lowering the NTU target somewhat to see whether this relaxes the fabrication problem. The exercise will illustrate an alternative use of the design charts.

8b Retain the earlier choice of 2.0 for L_{xc}/L_{ref} and fix number of tubes n_{Txc} at 100. (This promises a highly manageable 25 slots should things work out.) Using the left-hand graduations of the centre scale of the chart in Figure 15.3 locate corresponding to $L_{xc}/d_{xc} = 65$ [–].

9b Return to the chart in Figure 15.6, this time joining $L_{xc}/L_{ref} = 2.0$ of the right-hand scale to 65 of the L_{xc}/d_{xc} scale and extrapolate to the (un-graduated) turning scale. Join this latter point to the original value of 4.87×10^5 [–] on the N_{RE} scale and extrapolate to the NTU scale, finding $NTU = 0.75$ [–].

This lies *below* the 'balanced design' value – but *above* the figure for the GPU-3.

The re-design (Steps 8b–9b) took the writer approximately ½ min. This makes it a realistic proposition to spend all the time necessary to reconcile machining capability (number of slots) with target exchanger performance as measured by NTU.

15.4 Note of caution

The design sequence just illustrated places more reliance than most on the relevance to the Stirling gas processes of steady-flow heat transfer and flow correlations. 'More than most'

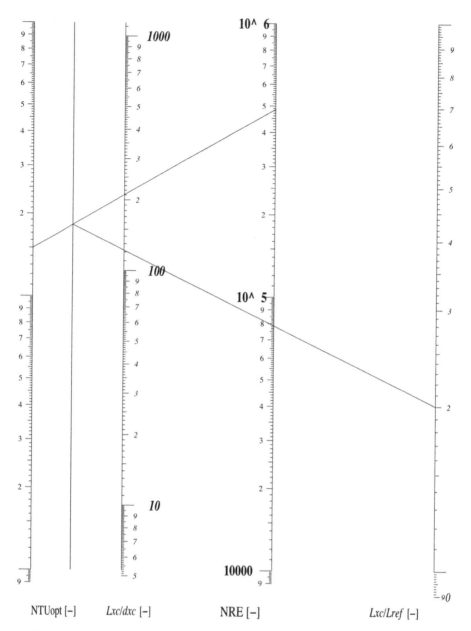

Figure 15.6 Optimum L_{xc}/d_{xc} in terms of N_{RE} and L_{xc}/L_{ref} for compression exchanger. *NTU* and N_{RE} scales can be connected, as can L_{xc}/d_{xc} and L_{xc}/L_{ref} (see text). No other connections allowed!

because the algebraic 'object function' which embodies them is submitted to *optimization*. The latter process – whether applied numerically of algebraically – is notorious for exploiting weaknesses in the object function: an inappropriate choice in setting up the formulation can result in an 'optimum' which is worse than an arbitrary design.

So where does this leave *Less Steam* as a design tool? It retains two potential values:

Codes of practice do not exist for any aspect of Stirling engine design. While this remains the case the designer might be wise to use all available design algorithms to 'bracket' a specification.

It can serve as a 'skeleton' into which relevant correlations can be plugged as and when available.

16

Heat transfer correlations – from the horse's mouth

16.1 The time has come

Concern has been voiced on many occasions as to the relevance of conventional, steady-flow heat transfer correlations. While no indication has been offered as to how the situation might be remedied, it must be self-apparent that the sole source of trustworthy data is the Stirling engine itself. The moment has arrived to explore how the 'Holy Grail' might be accessed.

The focus is multi-tube expansion exchangers as featured on General Motors' GPU-3, Philips' rhombic drive engines – the PD-46, and so on. Individual tubes in a bundle share common internal diameter d, all are of nominally the same length and all feature a 180-degree flow reversal at a point close to mid-length. This goes a long way to satisfying the fundamental *Geometric Similarity* requirement.

Figure 16.1 has been generated from data on Philips' 30-15 engine as reported by Hargreaves. Raw data were values for net thermal efficiency η [–] and shaft power P [W] for a range of charge pressure p_{ref} [Pa] and rpm [min^{-1}] for the 30-15 charged with H_2. The ratio P/η [W] is cycle-average heat rate Q'_{XE} [W] via the expansion exchanger, and permits calculation, for each p_{ref} and each rpm, of *specific thermal load* XQ_{XE} (derivation later):

$$XQ_{XE} = \frac{P/\eta}{\omega p_{ref} V_{sw}} \frac{N_T}{L_x/d} (\gamma - 1)/\gamma \tag{16.1}$$

In the figure, XQ_{XE} is plotted against exchanger characteristic Reynolds number $RE_{\omega XE}$:

$$RE_\omega = \frac{N_{SG}N_{MA}^2}{\delta_x f(T/T_C)} \frac{(L_x/V_{sw}^{1/3})^2}{L_x/d} \tag{16.2}$$

(The algebraic derivation of RE_ω is postponed in the interests of continuity.)

Stirling Cycle Engines: Inner Workings and Design, First Edition. Allan J Organ.
© 2014 John Wiley & Sons, Ltd. Published 2014 by John Wiley & Sons, Ltd.

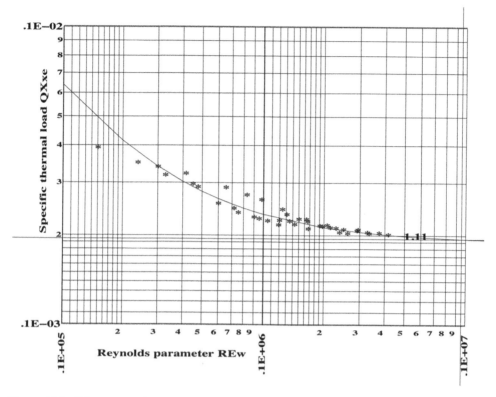

Figure 16.1 $XQ_{XE} - RE_\omega$ characteristic plotted from experimental measurements on Philips' 30-15 engine as reported by Hargreaves (1991)

The curve is fitted by:

$$XQ_{\omega XE} = C_1/RE_{\omega XE} + C_2 \tag{16.3}$$

$$C_1 = 4.0, \quad C_2 = 1.9\text{E-}04.$$

If XQ_{XE} and RE_ω are the appropriate choice of parameters, data for other multi-tube expansion exchangers may be expected to map similarly – but not necessarily to superimpose at this stage.

An engine equipped with a multi-tube exchanger comprising 20 parallel tubes differs fundamentally from one fitted with an exchanger of identical dead volume and free-flow area – but having, say, 40 parallel tubes: the pre-condition of (nominal) geometric similarity, essential for dynamic similarity, is lacking.

This resolves in the present case by correlating in terms of 'per tube' values: RE_ω values are unaltered, but specific thermal load per tube XQ_{TXE} takes over from XQ_{XE}.

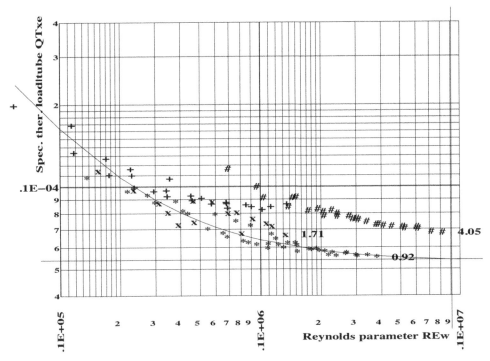

Figure 16.2   As for Figure 16.1, but with results superimposed from tests on GPU-3 and USS P-40

Key: * 30-15 (Hargreaves 1991); + GPU-3/H$_2$ (Percival 1974); × GPU-3/He (Tew et al. 1978); # United Stirling P-40 (Lundholm 1982). The numerical parameter is the nominal flush ratio (inverse of dimensionless dead space, δ_{xe})

Figure 16.2 includes points for the 30-15 engine (H$_2$-charged) re-plotted in this way, with counterpart points superimposed for: (a) General Motors' GPU-3 operating on H$_2$, (b) the same engine operating on He, and (c) the H$_2$-charged United Stirling P-40. GPU-3 results for H$_2$ (points '+') and He (points 'x') were acquired from different builds of the GPU-3, at different laboratories, using different instrumentation, at dates 4 years apart, and are documented to different format. Points # for the P-40 derive from tests at totally different facilities.

With this in mind, per-tube values of thermal load appear not to be heavily influenced by flush ratio. If dedicated tests eventually vindicate the XQ_{TXE} versus RE_ω relationship it will serve as a design tool capable of bypassing irrelevant steady-flow correlations. The possibility will be explored using the curve fitted to the Philips 30-15 engine. The latter had 36 tubes, so the coefficients of Equation 16.3, revised for the per-tube fit, are $C_1' = C_1/36 = 0.111$, $C_2' = C_2/36 = 5.39\text{E}-06$.

Combining curve-fit equation (Equation 16.3) and definitions:

$$\frac{P_T/\eta}{\omega p_{ref} V_{sw}} \frac{N_T}{4L_{xe}/d}(\gamma - 1)/\gamma = C_1' \frac{\delta_{xe}f(T/T_C)L_{xe}/d}{N_{SG}N_{MA}{}^2\left(L_{xe}/V_{sw}{}^{1/3}\right)^2} + C_2' \qquad (16.4)$$

In Equation 16.4 P_T is brake power per expansion exchanger tube. The equation is a quadratic in L_{xe}/d and of form $a(L_x/d)^2 + b(L_x/d) + c$, in which:

$$a = \frac{C_1' \delta_{xe} f(T/T_C)}{N_{SG} N_{MA}^2 (L_{xe}/V_{sw}^{1/3})^2}$$

$$b = C_2'$$

$$c = \frac{-P_T/\eta}{\omega p_{ref} V_{sw}} \frac{1}{4} N_T (\gamma - 1)/\gamma$$

The solution is:

$$L_{xe}/d = \frac{-b +/- \sqrt{\{b^2 - 4ac\}}}{2a} \tag{16.5}$$

Constant C_2' has a positive value making b positive, so the positive sign is selected.

16.2   Application to design

The designer will have chosen swept volume V_{sw}, charge pressure p_{ref}, working fluid and rated power P. The value of brake thermal efficiency η will be a target. Ideally the engine being designed should have the δ_{xe} of the engine(s) from which the correlation derives. (The numerical example which follows suggests that similarity of δ_{xe} values may not be critical.) The designer may fix a value for L_{xe} – which will almost certainly be influenced by space constraints, by the available method of fabrication, by the proposed heating provision, or by a combination. The design sequence requires an initial value of number of tubes n_{Txe}, which has to be a guess at this stage.

If a design is generated from an original specification there will be no means, short of building and testing, of checking the success – or otherwise – of the process. To furnish the desired check, the expansion exchanger of Philips' 400 hp/cylinder engine will be designed retrospectively. The exchanger is chosen for the fact that, 'though substantially larger than that of the 30-15, it is of the same, multi-tube construction involving the characteristic 180-degree flow reversal.

Table 16.1 lists data for the 400 hp engine used in back-calculating exchanger design. The data have to be supplemented by an estimate of thermal efficiency to allow brake power output to be converted to heat in. Hargreaves (1991) offers estimates of indicated thermal efficiency ($\sim$ 48.3%) – but does not specify a brake thermal value. The figure of 38.5% adopted here is a guess.

The hand calculator or the charts of Chapter 10 *Intrinsic similarity* are used to calculate $N_{SG} = 1.367E+10$ [–]; $N_{MA} = 0.0145$ [–]. If numerical values of δ_{xe} and L_{xe} cited in Table 16.1 are used, it will be possible to see if the design sequence returns internal diameter d_{xe} and number of tubes n_{Txe} of the 400 hp machine.

Processing the numbers yields $a = 0.112E-08$, $b = 0.527E-05$ (of course) and $c = -0.0586E-03$. Entering these values in Equation 16.5 gives $L_{xe}/d_{xe} = 108.48$ [–] the latter figure leading to $d_{xe} = 1090.0/108.48 = 10.048$ [mm] (c.f the 10.0 of the Philips specification).

Table 16.1 In generating Figures 16.1 and 16.2, test data on the Philips 30-15, the GM GPU-3 and the USS P-40 were processed over the pressure and *rpm* ranges indicated. Values for the P-40 are *per cylinder*. Data on Philips' 400 hp/cyl engine are from Hargreaves (1991) at p. 251

	Working fluid	30-15 H_2	GPU-3 H_2/He	P-40 H_2	400 hp/cyl He
V_{sw}	[cm³]	365	118	134	17 400
P_{ref}	[atm]	50–140	34–68	50–150	110
rpm	[min⁻¹]	250–2500	500–3600	1000–4000	452
L_{xe}	[mm]	465*	245	260	1090
d_{xe}	[mm]	5.5*	3.02	3.0	10.0
n_{Txe}	[–]	36*	40	18	49
$\delta_{xe} = V_{dxe}/V_{sw}$	[–]	1.089*	0.595	0.246	0.274

*Scaled from drawing(s) and/or photograph(s).

Dead space ratio δ_{xe} for the exchanger with n_{Txe} parallel tubes is ¼ $\pi d_{xe}^2 n_{Txe} L_{xe}/V_{sw} = 0.274$ for the present case. The number of tubes n_{Txe} is calculated as a real (rather than as an integer). Inverting to make n_{Txe} explicit gives $n_{Txe} = 49.0$ – versus the 49 of the Specification (Table 16.1)!

This is not quite the complete story: a first run of the sequence returns a value n_{Txe}^* of n_{Txe} improved relative to the initial, arbitrary choice. Repeating the computation with a new start value of ½($n_{Txe} + n_{Txe}^*$) leads to rapid convergence.

The extended horizontal and vertical lines superimposed on Figure 16.1 mark the RE_ω and XQ_{XE} values yielded by solution of Equation 16.5 for the 400 hp/cylinder engine. It is hardly surprising that the point of intersection for this extreme design lies beyond the experimental points from the 30-15 and GPU-3 tests. But it is a reassuring endorsement that the extrapolation generates almost exactly Philips' design from little more than a statement of brake power, thermal efficiency, and swept volume.

16.3   Rationale behind correlation parameters RE_ω and XQ_{XE}

Ideally the variation of the gas processes over a complete cycle would be defined in terms of the single numerical value of a parameter, or parameters – at first sight a tall order: at any given crank angle and gas path location (e.g., hot-side regenerator entrance) instantaneous mass rate m' [kg/s] differs by orders of magnitude between engines of different swept volume working fluid, *rpm*, and charge pressure.

On the other hand, all 'serious' Stirling engines operate with largely similar numerical values of the (dimensionless) parameters used to calculate ideal efficiency and ideal cycle work. These are temperature ratio, $N_T = T_E/T_C$, swept volume ratio $\kappa = V_C/V_E$, volume phase angle α and fractional dead-space distribution defined in terms of the various $\delta_x = V_{dx}/V_{sw}$. The numerical similarity is observed despite substantial variation in length, internal diameter, number and configuration of exchanger ducts.

Instead of pursuing the analysis in terms of m' (in absolute units, kg/s), define specific mass rate $\sigma' = d\sigma/d\varphi$:

$$\sigma' = m'RT_C/\omega p_{ref} V_{sw} [-] \tag{16.6}$$

The variation with crank angle φ of σ' at any given location (say, mid-point of the expansion exchanger) now differs little between engines of interest.

Hydraulic radius r_h for the cylindrical duct is $d/4$: Reynolds number Re for for this case is $Re = \varrho|u|d/\mu = |m'|d/\mu A_{ff}$.

Expressing m' in terms of σ' from Equation 16.6, substituting the ideal gas law $p/\varrho = RT$ and collecting terms yields:

$$Re = |\sigma'| \frac{N_{SG} N_{MA}{}^2}{\delta_x f(T/T_C)} \frac{\left(L_x/V_{SW}{}^{1/3}\right)^2}{L_x/d} \tag{16.7}$$

In Equation 16.7 δ_x is dimensionless dead-space ratio V_{dx}/V_{sw}, and $f(T/T_C)$ is the temperature correction (Sutherland law) for coefficient of dynamic viscosity μ. For the expansion exchanger $T/T_C = N_T$. For the compression exchanger $T/T_C =$ unity.

With cycle variation of σ' at given location known in advance, the numerical value of the remaining parameters in combination determines the cycle variation of Re. That combination is denoted RE_ω:

$$RE_\omega = \frac{N_{SG} N_{MA}{}^2}{\delta_x f(T/T_C)} \frac{\left(L_x/V_{sw}{}^{1/3}\right)^2}{L_x/d} \tag{16.8}$$

It is now necessary come up with an expression which similarly characterizes heat transfer per cycle: relate the per-cycle value of net exchanger heat rate, Q'_x, to parameters characterizing the gas processes occurring over a cycle internal to the exchanger. The numerical example is based on the expansion exchanger, where mean rate of heat supply to the outer surface of the exchanger is Q'_{XE}. It is assumed provisionally that it equates to mean rate of heat gained by the working fluid via cycle mean heat transfer coefficient h [W/m^{2}K] and corresponding mean value ΔT [K] of wall to gas temperature difference:

$$Q'_{XE} = h A_w \Delta T$$

The per-cycle value is acquired by dividing by $f =$ cps, where $f = \omega/2\pi$. Normalizing the result by the product of reference (mean) pressure p_{ref} and swept volume V_{sw}, viz, by $p_{ref} V_{sw}$ yields:

$$\frac{2\pi Q'_{XE}}{\omega p_{ref} V_{sw}} = \frac{2\pi h A_w}{\omega p_{ref} V_{sw}} \Delta T$$

There is a family resemblance to Finkelstein's 'dimensionless heat transfer coefficient' (Finkelstein 1960b).

Numerator and denominator of the right-hand side are multiplied by RT_C. For the ideal gas $R = c_p(\gamma - 1)/\gamma$. Substituting, cancelling 2π from both sides and recalling the definition of temperature ratio $N_T = T_E/T_C$:

$$\frac{Q'_{XE}}{\omega p_{ref} V_{sw}} = \gamma/(\gamma - 1)\frac{h}{c_p(\omega V_{sw}^{1/3}/\alpha_{ff})p_{ref}/RT_C}\frac{4L_x/d\,\underline{\Delta T/T_E}}{N_T} \qquad (16.9)$$

In Equation 16.9, α_{ff} is the ratio A_{ff}/A_{ref} of exchanger free-flow area A_{ff} to reference area $A_{ref} = V_{sw}^{2/3}$. Its inclusion notionally pro-rates peak piston face velocity $\omega V_{sw}^{1/3}$ to corresponding velocity in the exchanger. The term p_{ref}/RT_C has dimensions of density [kg/m^3], so the second group on the right-hand side has all the credentials of a Stanton number, St_ω.

A numerical value for the left-hand side can be calculated from experimental readings. Respective values of $\gamma/(\gamma - 1)$, of L_x/r_h and of N_T are known. Transferring known or knowable values to the left-hand side:

$$\frac{Q'_{XE}}{\omega p_{ref} V_{sw}}\frac{N_T}{L_x/d}(\gamma - 1)/\gamma = St_\omega\underline{\Delta T/T_E} \qquad (16.10)$$

If there is an analogy with the steady-flow case, St_ω must be a function of Reynolds number RE_ω. Moreover, $R_{e\omega}$ may be expected to determine temperature (and velocity) distributions. So, although it is not possible to calculate St_ω and $\Delta T/T_E$ separately, the *product* (and therefore the left-hand side) should be a function of RE_ω:

$$\frac{Q'_{XE}}{\omega p_{ref} V_{sw}}\frac{N_T}{L_x/d}(\gamma - 1)/\gamma = f(RE_\omega) \qquad (16.11)$$

Numerical values of the left-hand side are available from experiment.

Why a single term for cycle heat transfer in unsteady flow?

The First Law of Thermodynamics for element dx long of a duct having free-flow area A_{ff} and perimeter p_{erim} is:

$$h\Delta T p_{erim}dx = \partial/\partial t\{A_{ff}(c_v\varrho T)dx\} + \partial/\partial x\{A_{ff}(c_p\varrho uT)\}dx$$
$$\Delta T = (T_w - T) \qquad (16.12)$$

Assuming cyclic equilibrium, integrating the right-hand side over a cycle would yield zero for the first term – which can thus be ignored.

16.3.1   Corroboration from dimensional analysis

Dimensional analysis offers support: derivation starts from 12 variables having dimensions – Q'_{XE}, ω, p_{ref}, V_{sw}, and so on. Four dimensions are involved – M, L, T, and K. According to Buckingham's 'Pi' theorem, it should be possible to achieve a reduction of up to four on proceeding from variables having dimensions to corresponding dimensionless variables. One possible grouping resulting from application of the theorem is:

$$\frac{Q'_{XE}}{\omega p_{ref} V_{sw}} = f\{N_T, N_{SG}, N_{MA}, \delta_x, L_x/d, L_x/V_{sw}^{1/3}, \gamma\} \tag{16.13}$$

The dead volume ratio δ_x of Equation 16.8 is equivalent to the product of area ratio α_{ff} (of Equation 16.9) with $L_x/V_{sw}^{1/3}$, so the independent approaches have yielded the same inventory of dimensionless sub-groups.

17

Wire-mesh regenerator – 'back of envelope' sums

17.1 *Status quo*

Making regenerator design manageable means *characterizing* thermal performance in terms of the parameters of the operating environment – compressible, unsteady flow with cyclic variation of pressure.

Classic regenerator theory is a paragon of parametric formulation. However, even in Kolin's 1991 reworking (imaginative, beautifully illustrated but flawed) it has nothing to offer the designer of Stirling engines: what, after all, is the 'reduced period' Π of the regenerator installed in the MP1002CA; or its 'reduced length Λ'? Algebraic conversion of Π and Λ to the Stirling context is possible (if the assumptions of incompressible, frictionless flow can be tolerated), but another objection surfaces: extant temperature solutions in terms of Π and Λ have been achieved by algebraic sleight of hand which amounts to discarding the 'flush' phase. In the Stirling context the flush phase dominates.

The problem may not be beyond solution: Designing a Stirling regenerator does not call for the mathematician's generalized picture of transient thermal response. Attention can be confined to a narrow part of the potential operating envelope. Formulation then proceeds on the basis of prior knowledge of the analytical outcome of the pre-conditions.

17.2 Temperature swing

The perfect regenerator would deliver working fluid to the expansion exchanger at T_E, and to the compression exchanger at T_C that is, with no cyclic swing of local gas temperature T. In reality, a given location sees extremes of T twice per cycle. These connote two components: (a) excursion of matrix temperature T_w and (b) excursion ΔT $(= T - T_w)$ of gas temperature T relative to T_w. Varying pressure means that the two components are not in phase, but no specialist knowledge is required to predict that the ideal (zero overall swing) would call for a combination of infinite *thermal capacity ratio TCR* with infinite *NTU* (Number of Transfer Units).

Stirling Cycle Engines: Inner Workings and Design, First Edition. Allan J Organ.
© 2014 John Wiley & Sons, Ltd. Published 2014 by John Wiley & Sons, Ltd.

Under these (un-achievable) conditions, gas and matrix temperature distributions are linear and invariant between T_E and T_C, so that $\partial T/\partial x = \partial T_w/\partial x = (T_C - T_E)/L_r$. Running a sophisticated simulation with target values of NTU and TCR confirms that overall swing can be both minimal and symmetrical about the nominal gradient. The algebra to follow assumes that the only designs of interest are those which comply, that is, for which:

$$\partial T/\partial x = \partial T_w/\partial x \approx (T_C - T_E)/L_r \qquad (17.1)$$

A good basis for a new design is knowledge of what is achieved by benchmark technology.

17.2.1   Thermal capacity ratio

Provisionally overlooking thermal short (conduction of the matrix in the flow direction), and assuming that the specific heat c_w of the matrix material does not vary greatly with temperature, heat Q_w gained by the matrix during the cooling blow at cyclic equilibrium can be estimated:

$$Q_w = M_w c_w dT_w \qquad (17.2)$$

In Equation 17.2 dT_w is one-half of cycle temperature swing. Under the linearity assumption dT_w is measured at any point x along the axis, where $0 \leq x \leq L_r$.

A numerical value for M_w is available in terms of density ρ_w of the parent material, porosity $\P_v$ and dead volume V_{dr}:

$$M_w = \rho_w V_{dr}(1 - \P_v)/\P_v$$

An estimate of Q_w follows from an energy balance for the gas. The resulting ratio of matrix and gas temperature excursions will be approximately in the ratio of thermal capacities. That of the matrix is evidently $\rho_w c_w V_{dr}(1 - \P_v)/\P_v$. For flow at a constant rate, uniform density thermal capacity of the gas *per pass* would be $\rho c_p V_{dr} N_{FL}$, N_{FL} being flush ratio. There is no simple expression for N_{FL} – even for ρ constant and uniform – but the overriding term in the full calculation is inverse dead space ratio, that is, $N_{FL} \approx V_{dr}/V_{sw}^{-1} = \delta_r^{-1}$. Using a representative value of ρ, viz, $\rho = p_{ref}/RT_C$ and recalling the $R/c_p = (\gamma - 1)/\gamma$ yields thermal capacity ratio TCR:

$$TCR = N_{TCR}[(\gamma - 1)/\gamma]\delta_r(1 - \P_v)/\P_v \ [-] \qquad (17.3)$$

In Equation 17.3 N_{TCR} is a *nominal* thermal capacity ratio:

$$N_{TCR} = \rho_w c_w T_C/p_{ref} \ [-] \qquad (17.4)$$

Table 17.1 includes values of TCR back-calculated from published specifications of highly regarded engines. The figure of 15 for the P-40 reflects a massive δ_r of 1.298 for two regenerators per gas path. The high value of 20 for the MP1002CA follows from the modest charge pressure p_{ref} of 15 bar abs. and the correspondingly high N_{TCR}.

Table 17.1 *TCR* figures (Equation 17.3) for benchmark engines (truncated to nearest integer)

N_{RE}	Re	L_r/r_h	L_r/L_{ref}	TCR	Equipment
2.90E+06	268	1540	0.302	26	Philips 400hp/cylinder
6.13E+05	220	743	0.463	8	GPU-3
2.29E+05	195	665	0.946	20	Philips MP1002CA[†]
1.50E+06	274	1611	0.863	15	USS P-40
7.47E+05	248	1075	0.493	5	USS V-160
4.70E+05	116	916	0.475	17	Alison PD-46
2.69E+04	429	–	–	–	Ketek-1
4.43E+05	250	–	–	–	Kammerich
4.20E+04	781	197	1.462	–	Clapham 5cc[†]
4.00E+04	669	417	1.00	–	Stirling 1818[*]

[*]Reverse-engineered with the aid of computer simulation.
[†]Treated as screen stack having same r_h as wound-wire matrix actually installed.

The GPU-3 achieved its admirable peak thermal efficiency of 30% with a *TCR* of 8. The inference is that a design criterion of *TCR* > 10 would be a 'safe' one. Figure 17.1 is a nomogram giving, in terms of proposed (or available) volume porosity[1] $\P_v$ and dead volume ratio δ_r, the minimum value of N_{TCR} which will satisfy the provisional criterion. The designer may wish to start with δ_r of 0.4 (as per the 'standard specification' of Chapter 15, see Table 15.1). The choice at this stage is arbitrary. All other things being equal, however, low values connote high compression ratio and associated adiabatic losses, and may compromize later stages of the design sequence.

The nomogram of Figure 17.2 'knows' density and specific heat of the three specimen materials labelled at the left-hand scale. If the line connecting the chosen material with the N_{TCR} read from Figure 17.1 cuts the p_{ref} scale at a value *above* proposed charge pressure, *TCR* is greater than 10, as required. If it cuts *below*, *TCR* falls short in the ratio of cut value to proposed p_{ref}.

17.3 Aspects of flow design

The only prospect of characterizing flow behaviour and thermal response is by assuming one-dimensional – or 'slab – flow. The result will be credible if the flow itself can reasonably be considered one-dimensional – or at least axi-symmetric.

Free-flow area A_{ffxe} of the expansion exchanger tube bundle can be an order of magnitude less than A_{ffr} of the regenerator. For the P-40 $A_{ffxe}/A_{ffr} = 0.0328$, or 3.2%! There is incontrovertible evidence that flow from an individual tube of the expansion[2] exchanger emerges as a jet penetrating several layers of the gauze stack before dissipating. The phenomenon is evidently mitigated somewhat by provision of a plenum volume between tube exit and the first gauze of the stack (Figure 17.3). The matter is far from incidental: Feulner (2013) recounts experiments

---

[1] Aiming for $\P_v > 0.75$ minimizes likelihood of need for remedial action later when pumping penalty is estimated.
[2] The problem cannot be verified by the same means at the the compression end, but is likely to be less severe, as total inflow is distributed over a far larger number smaller tubes.

NTCR [–]

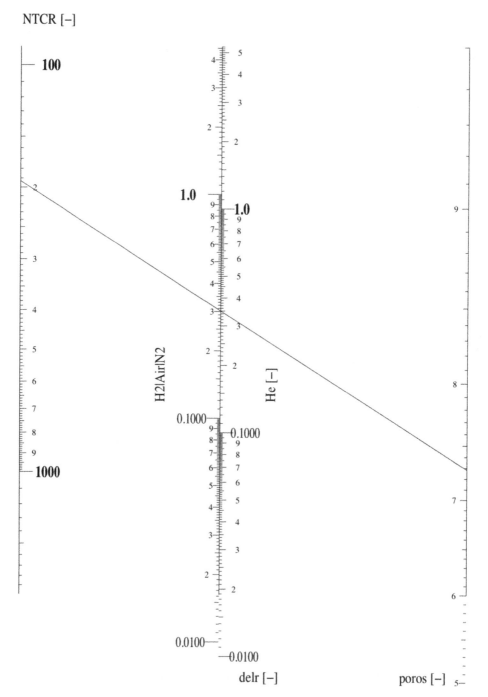

Figure 17.1 Nomogram for setting nominal thermal capacity ratio N_{TCR} in terms of dead space δ_r ratio and volume porosity $\P_v$. Scale values of $\P_v$ are *fractional, i.e.,* 0.6, 0.7, 0.8, and so on. The N_{TCR} scale reads *from top to bottom*

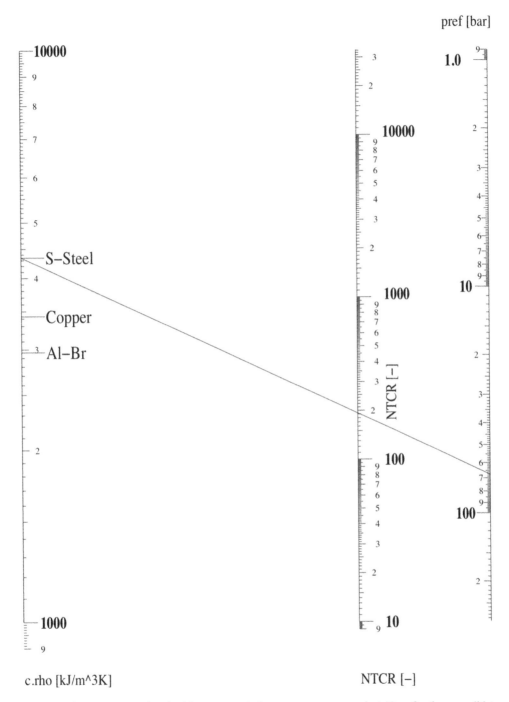

pref [bar]

c.rho [kJ/m^3K]   NTCR [−]

Figure 17.2   Nomogram for checking proposed charge pressure p_{ref} against N_{TCR} for three candidate matrix materials. The p_{ref} scale reads *from top downwards*

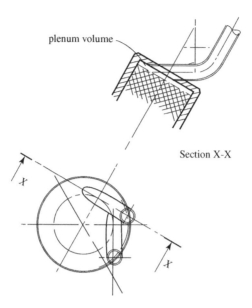

plenum volume

Section X-X

Figure 17.3   Measures for minimizing effects of 'jetting': plenum volume and provision for tangential entry

with engines which, with the exchanger end-plate in contact with the first gauze layer, failed to run but which, after introduction of a gap, operated satisfactorily. Stirling engine science would advance more rapidly if Peter Feulner could find time from his management responsibilities to publish an account of his impressive design and experimental work. Even when a plenum volume is provided, a degree of 'jetting' evidently persists, but can probably be reduced by introducing the flow with a tangential component, as suggested for a specimen tube in Figure 17.3.

With slab flow assumed, the algebraic start point is that for steady flow in the uniform-temperature matrix:

$$\Delta p = -\tfrac{1}{2}\rho u^2 C_f L_r / r_h \tag{17.5}$$

Substituting $\rho = p/RT$ for the ideal gas introduces local Mach number $Ma = |u|/\sqrt{(\gamma RT)}$ and allows re-writing:

$$\Delta p/p = -\tfrac{1}{2}\gamma Ma^2 C_f L_r / r_h \tag{17.5a}$$

Equation 17.5a introduces a further dimension to 'jetting', to which machines using heavy gases – air and N_2 – are vulnerable: Flow through the individual gauze aperture is subject to the effects of compressibility (choking) at values of approach Mach number, *Ma, an order of magnitude lower than the text-book criterion of Ma* ≥ 0.3 (the author's 2007 text)! At the area ratio cited above, the jet issues at speed u a factor of 30 times the nominal speed of flow through the gauze. This substantially increases the chances of approach *Ma* falling locally into the critical range. The individual jet penetrating with pressure gradient enhanced by incipient choking resolves into significant radial flows. Such components invalidate temperature and flow solutions based on the assumption of slab flow.

A search with the aid of Google Scholar suggests[3] that the matter of jetting has engaged other researchers, including Gedeon (1989) and Niu et al. (2003), underlining the reality and importance of the phenomenon.

Minimizing such irreversibilities involves the same priorities as aiming for one-dimensional flow. Assuming that all such measures are in place supports proceeding on the basis of Equation 17.5.

The term ρu is mass rate per unit area, viz m'/A_{ffr}. The obvious relationship between dead volume V_{dr}, A_{ffr} and L_r allows Equation 17.5 to be written:

$$\Delta p = -\tfrac{1}{2}m'^2 C_f L_r^3/(\rho r_h V_{\text{dr}}^2) \tag{17.6}$$

If the process of 'getting started' (Section 10.3) has led to a tentative swept volume V_{sw}, then V_{dr} follows immediately from the δ_r already indicated (Section 17.2) in terms of *TCR* requirement, viz V_{dr} [cm^3] $= \delta_r V_{\text{sw}}$[cm^3].

Volume V_{dr} is achieved as the product $L_r A_{\text{ffr}}$. Mass rate m' is independent of free-flow area, so with the numerical value of V_{dr} fixed by *TCR* requirement, pressure drop for given hydraulic radius r_h is proportional to L_r^3.

At first sight this leaves no doubt as to the design criterion for pumping loss: the shortest stack length for which the required *NTU* can be achieved.

Instantaneous pumping work dW_p across pressure difference Δp is $A_{\text{ff}}|u\Delta p|$. Evaluating over a cycle involves integrating between 0 and time t_{cyc} for a cycle, where $t_{\text{cyc}} = 2\pi/\omega$, and where ω in turn $= rpm \times 2\pi/60 \approx rpm/10$:

$$W_p = A_{\text{ff}} \int (|u\Delta p| dt \tag{17.7}$$

$$C_f = c/Re + d \tag{17.8}$$

An approximate fit to Kays' and London's (1964, at their Figure 7-9) C_f-Re correlation for $\P_v = 0.766$ is achieved with $c = 50$ and $d = 0.2$.

Substituting Equation 17.8 for C_f into Equation 17.7 for W_p, carrying out some cancelling and noting that dt can be replaced by $d\varphi/\omega$:

$$W_p = \tfrac{1}{2}L_r^2/(\omega r_h V_{\text{dr}}) \int \{(|u|m'^2(c/Re + d)/\rho\} d\varphi \tag{17.9}$$

---

[3] 'Suggests' because the sources have yet to be tracked down, and because Google Scholar, while invaluable, is not infallible: At the time of writing, the entry for the author's own 1997 title, *http://scholar.google.co.uk/scholar? hl=en&q=allan+organ+regenerator&btnG=&as_sdt=1%2C5&as_sdtp=*, attributes publication to Lavoisier (in fact it was MEP/PEP/Wiley), and gives the following helpful synopsis: *A first-response guide to medical problems that psychiatrists, as physicians, must be prepared to treat in their patients, both in-patient and out-patient. It may be searched by either specific diseases or symptoms, and covers internal medicine, neurology, chronic.*

If Equation 17.9 can be integrated over a cycle, dividing the result by $p_{ref}V_{sw}$ will reduce pumping loss to the same common denominator as a Beale modulus for a one-to-one comparison with useful cycle work.

The process will rely on expressing Re as a function of crank angle φ:

$$Re = \frac{4\rho|u|r_h}{\mu_0} = \frac{4|m'|r_h}{A_{ff}\mu_0} \tag{17.10}$$

In terms of the variables and parameters of intrinsic similarity:

$$Re = 4|\sigma'|\frac{(r_h/L_r)(L_r/L_{ref})}{\delta_r}N_{SG}N_{MA}{}^2 \tag{17.11}$$

Whether or not compressibility effects arise, the algebra will throw up local instantaneous Mach number $Ma = |u|/\sqrt{(\gamma RT)}$. Re-expressing in terms of similarity variables:

$$Ma = \frac{|\sigma'|\sqrt{\tau}}{\psi\alpha_{ff}\sqrt{\gamma}}N_{MA} \tag{17.12}$$

Equation 17.9 is evidently the sum of two separate integrals. Confining attention for the moment to the uniform-temperature case ($\tau = T/T_C$ – unity), substituting for Re and Ma into Equation 17.11 and re-arranging:

$$W_p/p_{ref}V_{sw} = Z_{p1} + Z_{p2} = \tfrac{1}{8}c/\delta_r\{(L_r/r_h)^2/N_{SG}\}\int(\sigma'/\psi)^2d\varphi$$
$$+ \tfrac{1}{2}d/\delta_r{}^2(L_r/L_{ref})^2(L_r/r_h)N_{MA}{}^2\int|\sigma'^3|/\psi^2d\varphi \tag{17.13}$$

Respective cycle histories of ψ and σ' from the Schmidt analysis are used for evaluating the integrals, which run between $\varphi = 0$ and $\varphi = 2\pi$. The process attracts all the usual objections to Schmidt, but cannot fail to be an improvement over the traditional assumptions of constant pressure and uniform, uni-directional mass rate. Numerical values are:

$$\int(\sigma'/\psi)^2d\varphi = CI1 = 0.0494 \quad \text{(viscous component)}$$

$$\int\sigma'^3/\psi^2d\varphi = CI2 = 0.0051 \quad \text{(inertia component)}$$

Having been calculated for the 'standard' specification of Table 15.1 they need not be re-computed, regardless of operating conditions (γ, N_{SG}, N_{MA}).

Figure 17.4 displays viscous component of cycle pumping loss Z_{p1} in terms of design variable L_r/r_h and operating parameter N_{SG}. Figure 17.5 shows inertia element Z_{p2} in terms of design variables L_r/r_h, L_r/L_{ref} and operating parameter N_{MA}. Net pumping loss $W_p/p_{ref}V_{sw}$, normalized to the same base as Beale number N_B, is estimated by adding the two components.

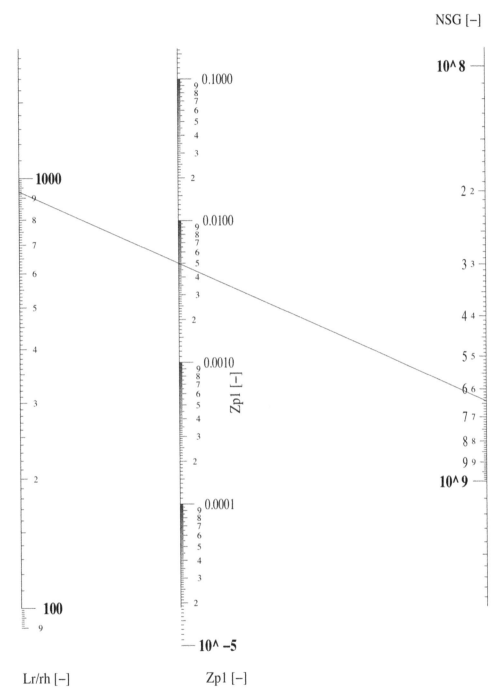

Figure 17.4 Viscous element of pumping loss Z_{pl} in terms of operating parameter N_{SG} and design variable L_r/r_h for 'standard' specification (Table 15.1)

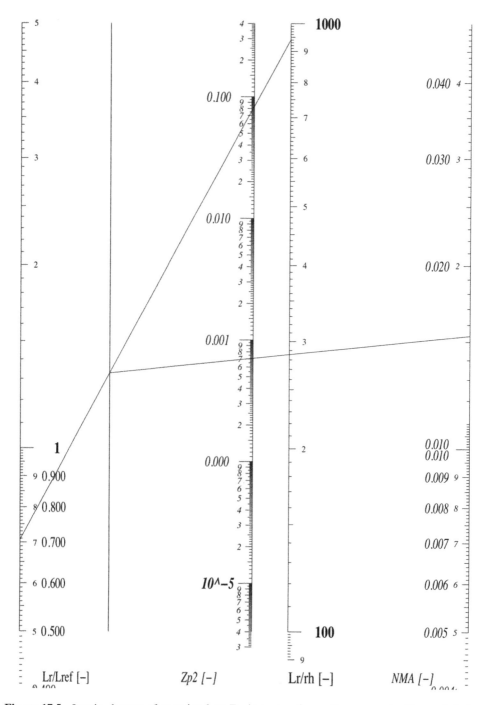

Figure 17.5 Inertia element of pumping loss Z_{p2} in terms of operating parameter N_{MA} and design variables L_r/r_h and r_h/L_r for 'standard' specification (Table 15.1). (*NB: A scale graduated in italics can be connected only to the other scale similarly graduated. Likewise, a scale annotated with upright characters can be connected only to the other such scale*)

17.4   A thumb-nail sketch of transient response

17.4.1   Rationalizations

A general solution of the regenerator problem is not required. The majority of the transient thermal response envelope is irrelevant. Good thermal recovery calls for the minimal temperature excursions already assumed.

Pressure p and local mass rates m' vary with crank angle as per the well-known Schmidt algebra.

Instantaneous Re is calculated from local, instantaneous m'. Instantaneous St follows in terms of Re from published steady-flow correlation appropriate to volume porosity.

An algebraic device is as important as, at first sight, it is unpromising. With reference to Figure 17.6:

Matrix dead volume $\delta_r = V_{dr}/V_{sw}$ is fixed while examining the effect of varying hydraulic radius and stack length. Unexpected benefits arise in relation to heat exchange intensity, which is that arising when a particle flows at velocity u relative to matrix temperature gradient $\partial T_w/\partial x$. In the absence of heat exchange, temperature difference $\Delta T = T - T_w$ when tracking the particle increases at rate $D\Delta T/dt = -u\partial T_w/\partial x$. Under present assumptions $-u\partial T_w/\partial x \approx -u(T_C - T_E)/L_r$. Varying L_r (without compensatory change in T_E and/or T_C) smacks of moving the goal posts. With

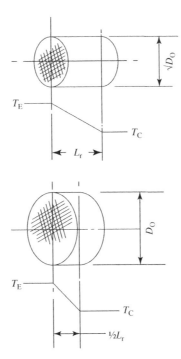

Figure 17.6   For given δ, the response to halved flow passage length L_r is a doubling of temperature gradient intensity – offset by doubled free-flow area A_{ff}. With r_h/L_r unaltered, Re is un-changed, as is NTU value and solution *independent of $\partial T_w/\partial x$!*

dead volume ratio δ_r fixed, however, halving L_r (thereby doubling $\partial T_w/\partial x$) is precisely offset (from the point of view of $NTU = StL_r/r_h$) by the doubling of free-flow area A_{ff} – and corresponding halving of u. (NB: expressed in terms of m', Reynolds number Re – and thus St – is independent of u – Equation 17.10).

Within the above constraints, thermal design reduces to selection of hydraulic radius r_h and length L_r. The algebra becomes particularly compact when formulated in terms of dimension-less variables (similarity variables) L_r/L_{ref} and r_h/L_r.

What ostensibly calls for return to first principles has been short-circuited here by the high temperature recover ratio required by the application, achievable only by a combination of high NTU and large thermal capacity ratio TCR^*. This amounts to a limiting case for compre-hensive regenerator solutions (e.g., those of the author, 1997, Chapter 7), which point to tem-perature distributions in both fluid and matrix of invariant, linear gradient $\partial T/\partial x \approx \partial T_w/\partial x \approx (T_C - T_E)/L_r$. With the exception of short distances close to $x = 0$ and to $x = L_r$, instantaneous temperature difference $\Delta T (= T - T_w)$ is independent of x.

This allows the cyclic variation in ΔT, T_w – and hence T – to be defined in simple algebra. Cycle variations of mass rates m_e' and m_c' at exit from expansion and compression exchangers, together with those of instantaneous pressure p (assumed uniform throughout the matrix) are those of the Schmidt algebra.

The start point is the standard thermodynamic relationship $ds = c_p DT/T - RDp/p$. The total differential operator D $(= \partial/\partial t + u\partial/\partial x)$ indicates that the relationship applies while following unit mass of gas. In the absence of viscous dissipation $ds = dq/T$. Recalling that $R = c_p(\gamma - 1)/\gamma$:

$$dq = c_p DT - c_p[(\gamma - 1)/\gamma]TDp/p$$

Heat rate per unit mass $q' = dq/dt$ can be expressed in terms of (variable) heat transfer coefficient h, and instantaneous local temperature difference $\Delta T = T - T_w$:

$$\frac{-hp_w \Delta T dt}{\rho c_p A_{ff}} = DT - [(\gamma - 1)/\gamma]TDp/p \qquad (17.14)$$

In Equation 17.14 p_w is wetted perimeter [m].

Being uniform with x at any instant, Dp/p can be written dp/p. DT is replaced by $DT - DT_w + DT_w$, and thus by $D\Delta T + DT_w$. By definition $DT_w/dt = \partial T_w/\partial t + u\partial T_w/\partial x$, where, by earlier hypothesis $\partial T_w/\partial x \approx (T_C - T_E)/L_r$ and where, in consequence of the same hypothesis $\partial T_w/\partial t = dT_w/dt$, so that $DT/dt \approx D\Delta T + \partial T_w/\partial t + u(T_C - T_E)/L_r$. Substituting into Equation 17.14:

$$\frac{-hp_w \Delta T}{\rho c_p A_{ff}} = D\Delta T/dt + dT_w/dt + u(T_C - T_E)/L_r - [(\gamma - 1)/\gamma]Tp^{-1}Dp/dt \quad (17.15)$$

Equation 17.15 is shaping up to be a first-order, total differential equation in ΔT of the form $D\Delta T/dt + P\Delta T = Q$. This will be susceptible to incremental solution over a succession of finite time increments Δt. The process will require a value for dT_w/dt at each step, dealt with by noting that the heat lost by a matrix element over time Δt is equal to that gained by the adjacent gas element, viz., to $hp_w dx\Delta T$:

$$dT_w/dt = \frac{hp_w\Delta T}{c_w\rho_w A_{ff}[(1 - ¶_v)/¶_v]} \qquad (17.16)$$

Recalling the definitions of u as $m'/\rho A_{ff}$, of $\sigma' = m'RT_C/\omega p_{ref}V_{sw}$, of Stanton number St (as $h/\rho|u|c_p$), of NTU (as StL_r/r_h) and of r_h as A_{ff}/p_w the reader may wish to recast Equation 17.15 in terms of similarity variables and parameters and verify:

$$P = |\sigma'|\tau NTU/(\psi\delta_r)$$
$$Q = -d\tau_w/d\varphi - \sigma'\tau(1 - N_T)/(\psi\delta_r) + [(\gamma - 1)/\gamma]\tau d\psi/\psi \qquad (17.17)$$
$$d\tau_w/d\varphi = |\sigma'|NTU\Delta\tau/TCR$$

The TCR defined at Equation 17.2 has turned up in context.

$$\Delta\tau_{\varphi+\Delta\varphi} = Q/P + (\Delta\tau\varphi - Q/P)e^{-P\Delta\varphi} \qquad (17.18)$$

As the integration process proceeds, a value of NTU is generated corresponding to varying mass rate $|\sigma'|$. An approximate fit to Kays' and London's correlations $Re - StPr^{2/3}$ for 'equivalent' stacked screens of volume porosity $¶_v = 0.766$ is $StPr^{2/3} = a/Re^b$, with $a = 0.958$ and $b = 0.429$:

$$NTU = StL_r/r_h = (0.958Re^{-0.429})L_r/r_h \qquad (17.19)$$

17.4.2 Specimen temperature solutions

Figure 17.7 gives solutions of Equations 17.17 and 17.18 for gas temperature T over a representative cycle for the MP1002CA and GPU-3 engines at respective rated operating conditions. The greater temperature swing of the GPU-3 is consistent with the lower thermal capacity ratio TCR (Table 17.1).

The solution algorithm can be run over a range of L_r/L_{ref} and L_r/r_h, noting peak temperature swing in each case. Figure 17.8 compares the outcome for the operating conditions of the MP1002CA and of the GPU-3.

Both sets of curves suggest: (a) that nothing (other than increased flow resistance) is to be gained by further reduction in the ratio r_h/L_r (i.e., by further increase in L_r/r_h and thereby in NTU), (b) that the minimum swing is set by thermal capacity ratio – higher in the MP1002CA than in the GPU-3 (Table 17.1), and (c) that there is the possibility of tackling the two components of temperature swing separately at the design stage.

The matter can be pursued by re-acquiring the curves with TCR set to infinity, as in Figure 17.9. The linear form of the log–log curves is confirmation that, for fixed dead volume ratio as assumed (see Section 17.4.1. *Rationalizations*), fractional temperature swing is a function only of the two parameters L_r/L_{ref} and L_r/r_h. The curves are already part-way to nomogram form. Figure 17.13 will exploit this by displaying the gas component of net swing as a nomogram for the 'standard' specification (Table 15.1).

There are various candidate matrix materials, but the square-weave wire gauze offers a unique attraction at the design stage: volume porosity $¶_v$, hydraulic radius r_h and aperture ratio α of the individual screen are completely specified by a mere two items of numerical

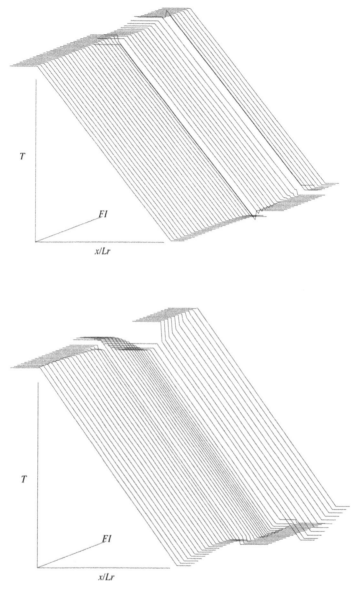

Figure 17.7   Gas temperature profiles for MP1002CA (upper) and GPU-3 (lower)

data: wire diameter d_w [mm] and mesh number m_w [mm^{-1}]. If the stack is close-packed, then per-screen values of $\P_v$ and α (and the ratio r_h/d_w) are approximately correct for the stack as a whole.

17.5   Wire diameter

If the mesh material is to be conventional, square weave, then $\P_v$ and r_h in combination now define wire diameter d_w and mesh number m_w. However, the process of arriving at the latter

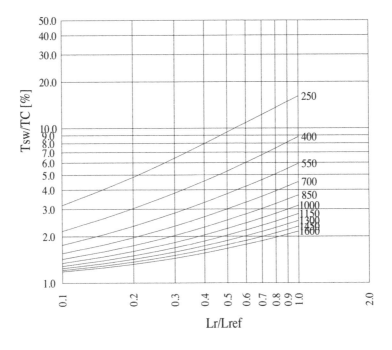

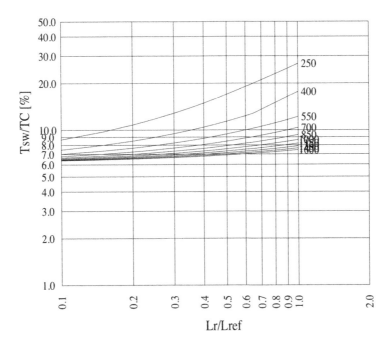

Figure 17.8 Net gas temperature swing (2 × amplitude) for MP1002CA (upper) and GPU-3 (lower)

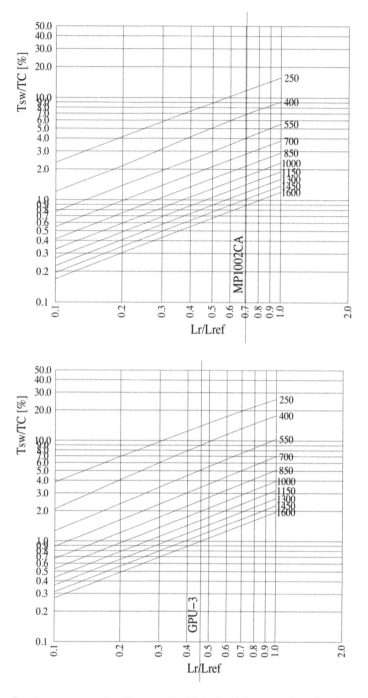

Figure 17.9 Gas temperature swing (2 × amplitude) isolated from matrix swing by setting $TCR =$ infinity. Upper set MP1002CA. Lower set GPU-3

pair of values has involved an assumption yet to be discussed – let alone justified – namely, that *the entire wire cross-section has been uniformly involved in the cyclic process of heat storage and release*. Put another way, the energy balance of Equation 17.16 assumes that a change in surface temperature T_w is instantaneously communicated, unattenuated as far as the centre-line. Miyabe et al. (1982) may have been the first to question the matter. They did so from the point of view of *thermal penetration depth*:

17.5.1   Thermal penetration depth

In the words of the originators:

> If the wire diameter d_w is too large, heat does not penetrate to the centre of the wire during the blow time. Thus some domain of the sectional area does not contribute to heat storage. If d_w is too small, heat penetrates to the centre of the wire before the blow time expires. In this case, heat storage volume is insufficient, although the total volume of the wire is used effectively.

The verbal assessment is backed up by their chart indicating d_w for optimum penetration in terms of operating conditions. The latter are expressed as Fourier modulus ($4\alpha t_{blow}/d_w^2$ in their notation, where α is thermal diffusivity [m^2/s]) and Biot modulus ($\frac{1}{2}d_w h/k_w$). Their model of transient thermal response thus couples heat rate through the convection film to diffusion below the wire surface. The expedient is routine in setting up Heisler charts (Chapman 1967, p. 135) of centre-line temperature history, but conflicts with the linearization targeted by the present approach (de-coupling the thermal response of the gas from that of the matrix).

The priority can be restored by considering the diffusion equation for the infinite cylinder:

$$\partial^2 T/\partial r^2 + r^{-1}\partial T/\partial r = \alpha^{-1}\partial T/\partial t \tag{17.20}$$

Analytical solutions (Schneider 1955, Chapman, *op. cit.*) are available in terms of Bessel functions, but numerical solution is quicker and more flexible: With subscript j to indicate radial location, and i to indicate location in the time frame the differential coefficients are 'discretized':

$$\partial^2 T/\partial r^2|_{i,j} \approx \frac{T_{i,j+1} - 2T_{i,j} + T_{i,j-1}}{\Delta r^2}$$

$$\partial T/\partial t|_{i,j} \approx \frac{T_{i+1,j} - T_{i-1,j}}{2\Delta t}$$

Figure 17.10 shows the computed temperature envelope for a 0.04 mm diameter wire of stainless steel to AISI 316 ($\alpha = 4 \times 10^{-6}$[m^2/s]) nominally at 300 K and subjected to simple-harmonic temperature disturbance at the surface of amplitude 10 °C. Amplitude is attenuated to approximately one third at the axis. (The disturbance frequency causing this degree of attenuation will be given shortly in terms of *rpm*).

Generating a chart of centre-line temperature history in the Heisler tradition involves impos-ing and maintaining a 'one-off' step change $T_s - T_0$ at the periphery, where T_0 is the initial, uniform temperature. Response at the centre-line is monitored as a function of time t and the 'error' relative to the eventual equilibrium state plotted as a fraction (or percentage)

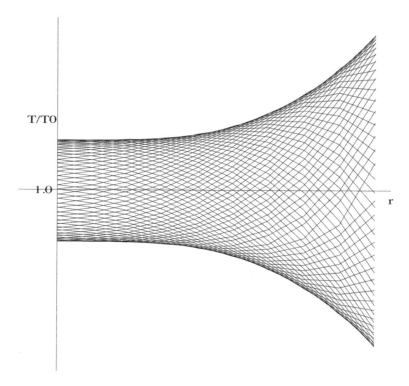

Figure 17.10   Envelope of simple-harmonic temperature disturbance propagating from periphery (right-hand side) to axis of infinite cylinder. Parent material: AISI 316 stainless steel ($\alpha = 4 \times 10^{-6}[\text{m}^2/\text{s}]$). Diameter $d_\text{w} = 0.04$ mm

ε_T of $T_\text{s} - T_0$. In the context of the Stirling engine, time t is meaningless by itself until related to crank angle. Achieving this via the self-evident $\varphi = \omega t$ throws up a parameter – a modified Fourier number $N_\text{F} = \alpha\omega/(\tfrac{1}{2}d_\text{w})^2$. The completed chart follows as Figure 17.11.

Miyabe et al. propose trial-and-error search for d_w giving a centre-line 'error' of between 0 and 5% (their $1.0 > T_\text{r} > 0.95$). The former figure (100%) being unattainable, Figure 17.11 is entered at the 5% point of the left-hand scale. A horizontal line extended right-wards intersects curves of decreasing N_F at values of elapsed crank angle for which centre-line 'error' ε_T is 5%.

If the designer hopes for thermal saturation corresponding to 95% of surface temperature disturbance within one degree of crankshaft rotation, then N_F of 32.60 is required. Inverting the expression for N_F:

$$\tfrac{1}{2}d_\text{w} = \sqrt{\{\alpha/(N_\text{F}\omega)\}} \quad [\text{m}] \tag{17.21}$$

Crankshaft *rpm* of 1500 convert to $\omega \approx 150$. On the assumption of stainless-steel wire to AISI 316 ($\alpha = 4.53 \times 10^{-6}[\text{m}^2/\text{s}]$) $d_\text{w} = 2\sqrt{\{4.53 \times 10^{-6}/(32.60 \times 150)\}} = 0.061$ mm, or 0.0024 inch. To satisfy the same criteria, the USS P-40 engine at 4000 *rpm* would require wire

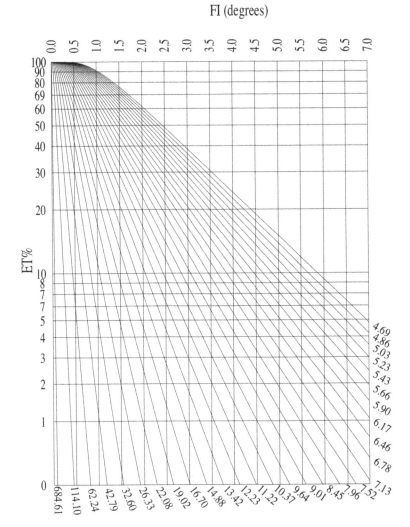

Figure 17.11   Centre-line temperature 'error' $\varepsilon_T\%$ as a function of elapsed crank angle φ with Fourier modulus as parameter

diameter decreased in the ratio $\sqrt{1500/4000}$, that is, $d_w = 0.037$ mm, or 0.0015 inch. Both figures lie in the range of diameters commonly specified and commercially available.

17.5.2   Specifying the wire mesh

The designer following the guidelines of this chapter now faces a choice between (a) converting hydraulic radius r_h from Section 17.4 to d_w (and m_w) via porosity $\P_v$ or (b) accepting the d_w from consideration of thermal penetration depth. Pursuing either option involves making arbitrary assumptions. Coin-tossing is probably minimized by taking option (a) while reserving (b) for corroboration.

The complete specification (d_w and m_w) of the square-weave gauze can be read in terms of r_h (recently calculated) and required volume porosity $\P_v$ from Figures 17.12 to 17.16.

All three design variables r_h, $\P_v$ and aperture ratio α_w are functions of the dimensionless product $d_w m_w$ [−], which serves as geometric similarity parameter: two screens, each square-woven in different wire diameters, have geometrically similar flow passages (and share the same $\P_v$ and same α_w) provided only that $d_w m_w$ has the same numerical value in both cases.

$$\text{volume porosity } \P_v \approx 1 - \tfrac{1}{4}\pi d_w m_w \sqrt{\{1 + (d_w m_w)^2\}} \text{ [−]} \qquad (17.22)$$

$$\approx 1 - \tfrac{1}{4}\pi d_w m_w \text{ [−]} \qquad (17.22a)$$

The linear-scale variable of pressure drop and heat transfer sums is hydraulic radius r_h:

$$r_h/d_w = \tfrac{1}{4}\P_v/(1 - \P_v) \qquad (17.23)$$

An upper limit of $1/\sqrt{3}$ is imposed on $d_w m_w$ by geometric considerations.

The charts which follow are in convenience units [mm]. (**NB: Use of r_h in flow and heat transfer calculations requires a numerical value in metres [m].**)

Compressibility effects are a function of (dimensionless) aperture ratio, α_w [−]:

$$\alpha_w = (1 - d_w m_w)^2 \qquad (17.24)$$

The expression for α employed by Pinker and Herbert (1967) is more sophisticated, but remains a function of $d_w m_w$ only.

17.6   More on intrinsic similarity

In respect of working fluids, charge pressures and *rpm*, and of gas path geometry (small number of long, large-diameter expansion exchanger tubes of the former versus large number of short, narrow slots in the latter) the GPU-3 and MP1002CA could hardly be more disparate. The close similarity of respective gas temperature swing characteristics (Figure 17.9) confirms a high degree of inevitable *intrinsic similarity*.

Respective L_r/L_{ref} values are marked, and intersect corresponding L_r/r_h at about 3%T_C or 9 °C for the MP1002CA, and 3.5%T_C or 10.5 °C for the GPU-3. (Both designs originated with Philips!)

The result gives some confidence in including Figure 17.15 to aid selection of L_r/r_h in terms of L_r/L_{ref} for candidate designs to the outline 'standard' specification of Table 15.1. The oblique solution line through $T_{sw} = 2\%T_C$ is an arbitrary choice. Entering the chart with the L_r/L_{ref} and L_r/r_h values of the MP1002CA suggests a temperature swing reduced to 1.5% (from the 30% larger regenerator provision corresponding to $\delta_r = 0.4$ of the 'standard' specification).

Sufficient background is now in place to justify a return to the temperature envelope of Figure 17.10: achieving this degree of attenuation of the surface temperature disturbance required a crank-shaft speed of 1 000 000 (one million) *rpm*! If the figure is counter-intuitive, imagine a wire of $d_w = 0.04$ mm initially at uniform T_0 subjected to a temperature disturbance

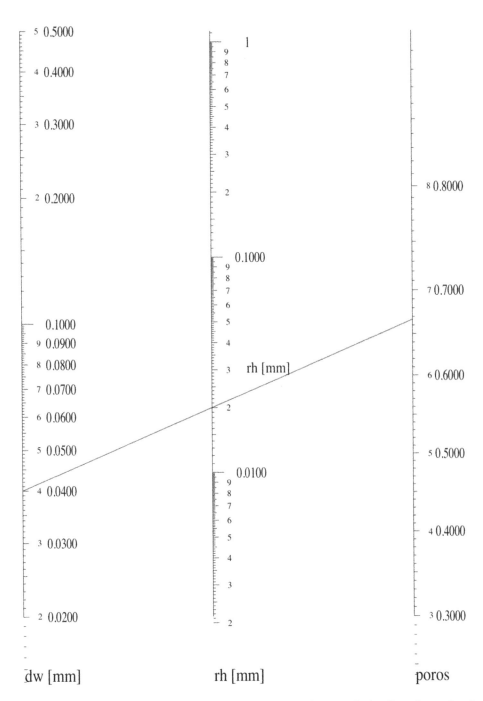

Figure 17.12 Chart for inter-conversion of wire diameter d_w [mm], hydraulic radius r_h [mm], and volume porosity $\P_v$ [–]. NB: check that product $d_w m_w$ does not exceed $1/\sqrt{3}$

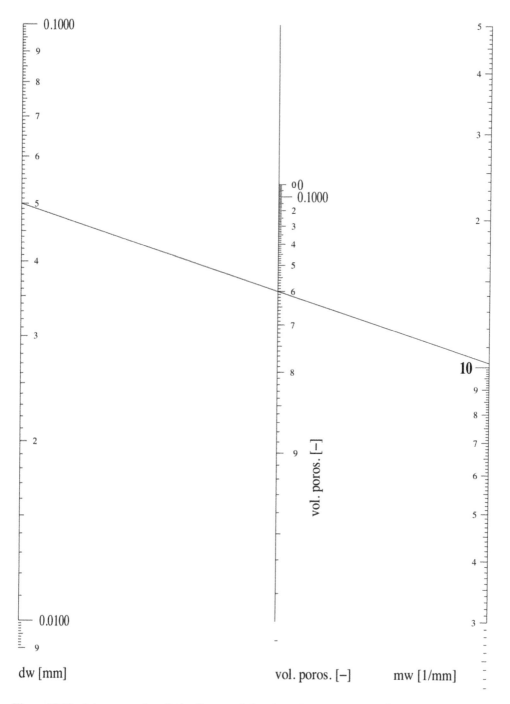

dw [mm]   vol. poros. [−]   mw [1/mm]

Figure 17.13   Inter-conversion of wire diameter d_w [mm], mesh number m_w [mm^{-1}] and volume porosity $\P_v$ [−]. NB: Commercial availability appears to be limited to $\P_v$ given by $d_w m_w > 0.3$

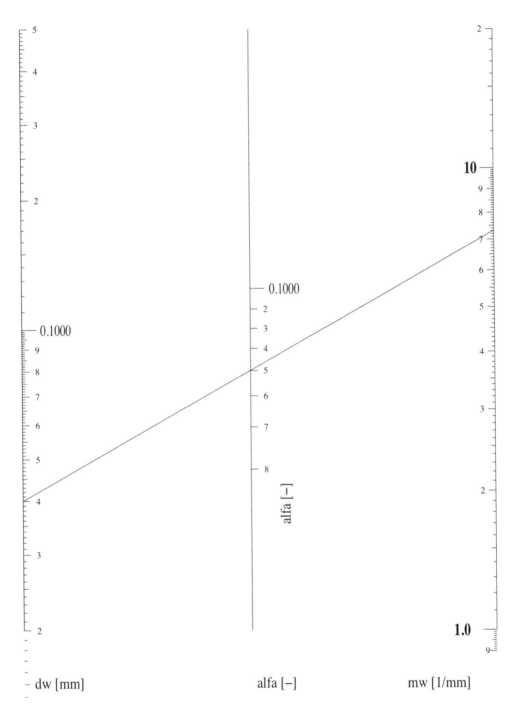

Figure 17.14 Inter-conversion of wire diameter d_w [mm], mesh number m_w [mm^{-1}] and aperture ratio α_w [–] **NB: check that product $d_w m_w$ does not exceed $1/\sqrt{3}$**

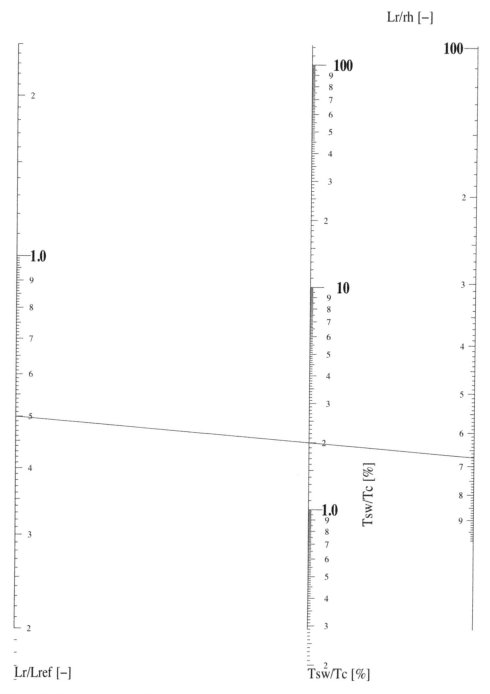

Figure 17.15   Nomogram equivalent of Figure 15.9 (temperature swing – gas relative to matrix) constructed for 'standard' specification of Table 15.1

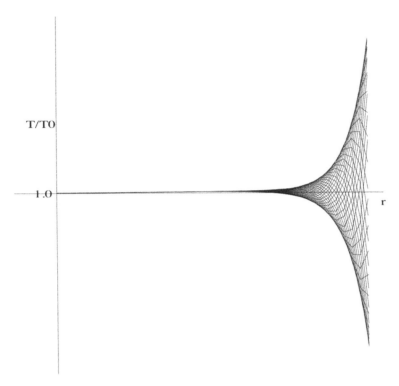

T/TO

1.0

r

Figure 17.16 Envelope of temperature wave computed for cylindrical bar of $d_w = 9.5$ mm (⅜ inch) having diffusivity of AISI 316 stainless steel. Frequency of surface disturbance equivalent to 1000 *rpm*

at the outer surface of 10 °C. To facilitate the arithmetic, the initial temperature profile is linear from outer diameter to the axis, meaning that $\partial^2 T / \partial r^2 = $ zero, and $\partial T / \partial r = 10.0/0.04 \times 10^{-3}$, or an impressive 250 000 [°C/m].

According to Equation 17.23, initial rate of re-adjustment $\partial T / \partial t$ of the internal temperature profile under this assumption is 4×10^{-6}[m²/s] $\times 250 \times 10^3$[°C/m]$/0.02 \times 10^{-3}$[m], or $50\,000$ °C/s. Time for a complete cycle at 10^6 *rpm* is $60/10^6$ s, suggesting a temperature excursion per cycle of $50\,000 \times 6.0 \times 10^{-5} = 3$ °C, or 30% of the magnitude of the disturbance. The result of the crude arithmetic is not inconsistent with Figure 17.15.

Figure 17.10 suggests substantial thermal storage capacity even under unfavourable conditions. This calls into question the need for complete penetration. A view may emerge on looking back at regenerator heat transfer work which has become classic: Kays and London (1964) achieved $StPr^{2/3}$ – *Re* correlations in the high Reynolds number range by use of rods of d_w up to 9.5 mm (⅜ inch). Assuming the rod material to have had diffusivity α of AISI 316 stainless steel, Figure 17.16 shows the temperature penetration envelope at 1000 *rpm*. Thermal soakage even at these modest *rpm* was evidently a fraction of the potential of the cross-section. The transient analysis used by Kays and London to back-calculate film coefficient h, and hence the $StPr^{2/3}$ – *Re* correlations, was Schumann's 'initial blow' solution of 1929. This is formulated on the assumption that temperature within the solid is uniform at

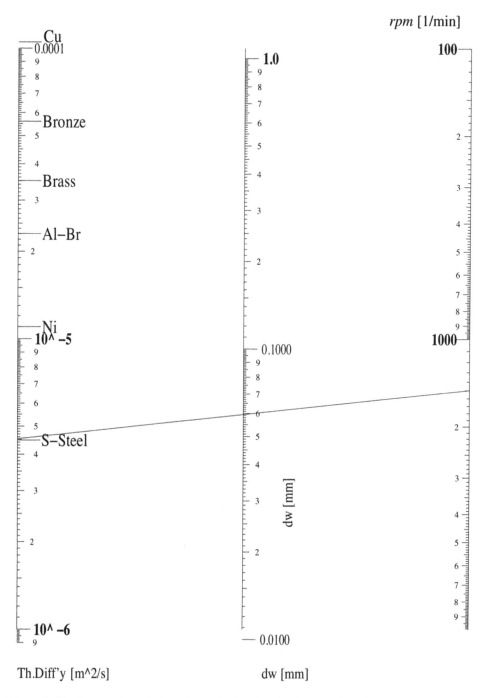

Figure 17.17 Nomogram equivalent of Equation 17.21a: wire diameter d_w to give 95% thermal soakage within one degree of crankshaft rotation (Figure 17.11). Specimen solution (oblique line) corresponds to numerical example of Section 17.5.1

any instant. The effect of the anomaly on the published $StPr^{2/3} - Re$ correlations will probably never be known. However, the key to banishing misgivings from future work is now self-evident: it involves achieving the N_F which yielded the d_w corresponding to the 95% thermal soakage of the numerical example at para. 17.5.1, viz $N_F \geq 32.6$ [−]. Noting that $\omega = 2\pi rpm/60$ and taking the numerical constant outside the root, Equation 17.21 for d_w becomes:

$$d_w \leq 1.082\sqrt{\{\alpha/rpm\}} \text{ [m]} \tag{17.21a}$$

Figure 17.17 is in 'convenience' units [rpm, mm] and allows quick evaluation of inequality Equation 17.21a for parent materials of the wire mesh.

Hargreaves' otherwise authoritative account[4] of 1991 offers the criterion $\frac{1}{2}d_w \ll \alpha/\omega$. This is dimensionally incorrect. His accompanying numerical example reflects recognition of the missing square root, but introduces a numerical error of $\sqrt{2}$.

---

[4]Hargreaves was editor of *Philips Technical Review*, which carried the enviable privilege of access to high-quality photographic illustrations, professionally drawn diagrams – and, of course, to Philips' personnel. It is not clear whether Hargreaves contributed to the research and development, or whether his book is merely an (impressive and largely accurate) account of the work of others.

18

Son of Schmidt

18.1 Situations vacant

There is an urgent need to replace the Schmidt analysis with something more helpful to the designer.

In the hands of Finkelstein (1960a) the algebra per se is seductively elegant. Within its own terms of reference it achieves everything a symbolic model could aspire to: the parametric formulation succinctly *characterizes* the thermodynamic system which it idealizes.

But that idealization is so far from the realities of a Stirling engine as to be treacherous. The problem lies squarely with the arbitrary and untenable assumption of isothermal working spaces, under which the cylinders take the entire thermal load, leaving the exchangers with nothing to do. Chapters 7 and 8 have gone a long way to confirming that the processes in the variable-volume spaces are closer to adiabatic than to isothermal. By any logic this leaves the exchangers shouldering the net thermal load. They need designing accordingly.

It is unclear whether this is the background to previous attempts to supersede Schmidt: Setting the dimensionless heat transfer coefficient to zero in the legendary *Generalized Thermodynamic Analysis* amounts to achieving adiabatic conditions in the working spaces (Finkelstein 1960). However, the perfect regenerator and heat exchangers mask the penalty of the increased thermal load. The adiabatic analysis applied by Urieli and Berchowitz to operating conditions of General Motors' GPU-3 engine similarly compensates increased thermal load with arbitrary exchanger performance. The author's own exercise in optimization (Organ 1992) incorporated the same idealization – and accordingly identifies false optima of volume ratio κ and phase angle α.

A disincentive to discarding Schmidt is its undeniable elegance. Nevertheless, a finely turned-out piece of algebra which subverts reality in all respects that matter simply has to go. But what then?

Most people evidently have no trouble identifying a numerical algorithm with which to beat a differential equation into submission. This, together with the fact that coding for the digital computer has a vastly greater following than applied mathematics, makes the numerical 'solution' the line of least resistance. But the analyst is not yet out of a job.

Stirling Cycle Engines: Inner Workings and Design, First Edition. Allan J Organ.
© 2014 John Wiley & Sons, Ltd. Published 2014 by John Wiley & Sons, Ltd.

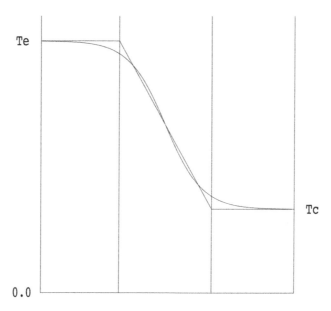

Figure 18.1   Wall surface temperature distribution generated from Equation 18.1

18.2   Analytical opportunities waiting to be explored

The analytical approach dislikes discontinuities[1]. The gas path of the typical Stirling engine has several changes in free-flow area – but a gas occupies *volume*, and cumulative volume V, reckoned from a piston face, is continuous. The algebraic transformation from x as independent variable to V could hardly be more straightforward.

Another discontinuity which, outside the context of Schmidt, threatens problems, occurs in the temperature distribution across heat exchangers and regenerator – traditionally depicted piece-wise linear as in Figure 18.1. A distribution which is equally – or more – plausible is available in the hyperbolic function $y = \tanh(x)$, which varies from -1 at $x = -\infty$ to $+1$ at $x = +\infty$. Amplitude is unity, whereas that of the target temperature distribution is $\frac{1}{2}(T_E - T_C)$. Moreover, the sign of y is the inverse of that required by the opposed piston layout. The relevant adjustment is achieved as $y = \frac{1}{2}(T_E + T_C) - \frac{1}{2}(T_E - T_C)\tanh(x)$. Addition of a constant to x shifts the point of inflexion (maximum gradient) to the regenerator mid-plane.

Finally:

$$T_w(x) = \tfrac{1}{2}(T_E + T_C) - \tfrac{1}{2}(T_E - T_C)\tanh(x') \tag{18.1}$$

$$x' = -a(L_e + L_{xe} + \tfrac{1}{2}L_r)$$

In Equation 11.1 a is a numerical multiplier which can be varied to 'stretch' the distribution horizontally to the desired fit. The L_e, L_{xe} and L_r are the physical lengths of expansion cylinder,

---

[1] An arguable exception is linear wave theory which, thanks to Lighthill (1975), thrives on them.

expansion exchanger and regenerator respectively. They may be substituted by respective volume values with corresponding adjustment to a. Local gradient follows in terms of the standard differential of tanh, viz sech^2:

$$\mathrm{d}T_{\mathrm{w}}/\mathrm{d}x = -\tfrac{1}{2}(T_{\mathrm{E}} - T_{\mathrm{C}})\,\mathrm{sech}^2(x') \tag{18.2}$$

Coding of Equations 18.1 and 18.2 as functions for computer implementation is a routine matter. Figure 18.1 is a plot of Equation 18.1 for $a' = 2.5$.

Progress on the analytical front awaits the attention of a dedicated specialist. The interim compromise has five requirements:

> It should not grow into a simulation: the real cycle is a thermodynamic pudding of competing processes. The perfect simulation is the same – making it impossible to distinguish individual cause and effect.

> Like Schmidt, it should ignore the effects of viscosity to highlight heat/work interactions.

> It should be formulated in terms of the parameters of the design and operating conditions of the real engine.

> Local, instantaneous heat exchange intensity should reflect respective flow conditions.

> Output display should be readily identified with – and verifiable against – the starting assumptions on which it is based.

The balance of this chapter will explore the mileage available from the assumption that mass in a given exchanger element, while *unsteady* with respect to the independent variable φ (crankshaft location), is *uniform* with respect to the other independent variable, location x.

18.3   Heat exchange – arbitrary wall temperature gradient

In steady flow at uniform pressure:

$$h \cdot p_{\mathrm{w}}\{T_{\mathrm{w}}(x) - T(x)\}\mathrm{d}x = \rho u c_{\mathrm{p}} A_{\mathrm{ff}}\mathrm{d}x\, \mathrm{D}T/\mathrm{d}t$$

$$\mathrm{D}T/\mathrm{d}t = \partial T/\partial t + u\partial T/\partial x$$

In steady flow $\partial T/\partial t = 0$. Cancelling the d$x$, adding $\partial T_{\mathrm{w}}/\partial x$ to the right-hand side – and subtracting again:

$$\frac{hp_{\mathrm{w}}}{\rho u c_{\mathrm{p}} A_{\mathrm{ff}}}\, \{T_{\mathrm{w}}(x) - T(x)\} = \partial T/\partial x - \partial T_{\mathrm{w}}/\partial x + \partial T_{\mathrm{w}}/\partial x$$

The dimensionless ratio $h/\rho u c_p$ [–] is Stanton number St, while A_{ff}/p_w is hydraulic radius r_h [m]. Abbreviating temperature difference (or 'error') $T - T_w$ to ε_T leaves:

$$-(St/r_h)\varepsilon_T = \partial\varepsilon_T/\partial x + \partial T_w/\partial x$$

Replacing the partial differentials by totals (steady flow) and re-arranging:

$$\frac{d\varepsilon_T}{(St/r_h)\varepsilon_T + dT_w/dx} = -dx \tag{18.3}$$

If wall temperature gradient dT_w/dx is uniform over length L_r of the regenerator (e.g., $dT_w/dx = (T_C - T_E)/L_r$ then Equation 18.3 is of the form:

$$\frac{dy}{ay + b} = -dx$$

The indefinite integral of the left-hand side is the natural logarithm of the denominator. On integrating between 0 and x, noting that $StL_r/r_h = NTU$ and converting to dimensionless variables $\lambda = x/L_r$ and $\tau = T/T_C$, temperature error $\varepsilon_{\tau\lambda}$ at location λ ($0 < \lambda < 1.0$) is expressed in terms of the inlet value $\varepsilon_{\tau i}$ as:

$$\varepsilon_{\tau\lambda} = -(d\tau_w/d\lambda)/NTU + \{\varepsilon_{\tau i} + (d\tau_w/d\lambda)/NTU\}e^{-\lambda NTU} \tag{18.4}$$

For the special case already anticipated whereby $dT_w/dx = (T_C - T_E)/L_r$:

$$\varepsilon_{\tau\lambda} = -(1 - N_T)/NTU + \{\varepsilon_{\tau i} + (1 - N_T)/NTU\}e^{-\lambda NTU} \tag{18.4a}$$

Absolute (dimensionless) temperature distribution follows immediately from the definition of ε_τ as $\tau - \tau_w$:

$$\tau_\lambda = \tau_w - (1 - N_T)/NTU + \{\varepsilon_{\tau i} + (1 - N_T)/NTU\}e^{-\lambda NTU} \tag{18.5}$$

It is evident that inlet 'error' can be positive, negative, or zero. Figure 18.2 illustrate the three possibilities over a range of NTU.

The first matter of note is that the solutions do not agree with any of those in Jakob's wide-ranging exposition of 1957.

At values of NTU at or below 2.0 (Figure 18.2a, b) calculated temperature recovery is inadequate for efficient functioning: The value is of the order of NTU back-calculated for the regenerative annulus of the 'hot air' engine.

Inlet temperatures illustrated are factors of 1.1, unity and 0.9 of respective nominal end temperatures T_E and T_E. It appears that $NTU > 2.5$ are necessary to iron out resulting outlet temperature differences (Figure 18.2b). $e^{-NTU} \approx 1/(1 + NTU)$ and outlet temperature $\tau_{\lambda=1}$ for flow left-to-right at $x/L_r = 1$ are then adequately predicted by:

$$(\tau - \tau_w)_{\lambda=1} \approx \{(1 - N_T) + \varepsilon_{\tau i}\}/(1 + NTU) \tag{18.6}$$

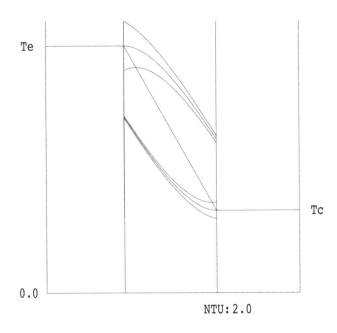

NTU: 2.0

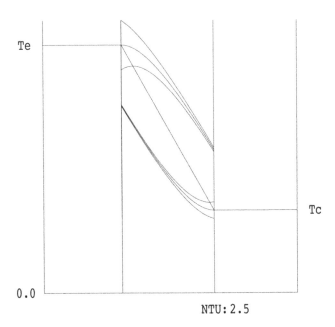

NTU: 2.5

Figure 18.2a   Quasi steady-flow temperature solutions (Equation 18.5) for low *NTU*

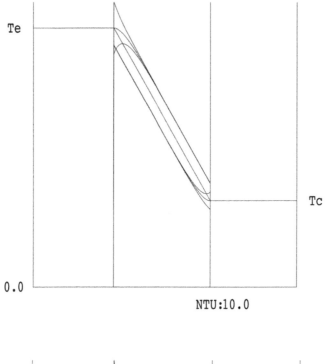

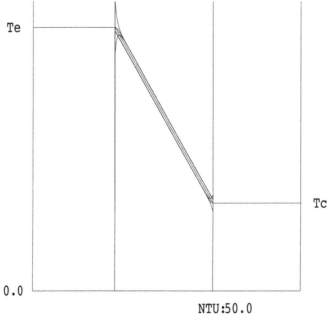

Figure 18.2b   Quasi steady-flow temperature solutions (Equation 18.5) for *NTU* affording viable temperature recovery

Equation 18.6 is also instantly recognizable as the solution for all values of NTU for the case of the uniform-temperature exchanger ($dT_w/dx = d\tau_w/d\lambda = 1 - N_T = $ zero).

Solutions at Figure 18.2d are commensurate with early claims by Philips of temperature recovery ratios of 90%. At 50, corresponding NTU tallies with values back-computed for Philips-type wire-screen regenerators in H_2- and He-charged engines.

Regenerator temperature distribution is plotted by substituting for λ ($= x/L_r$) in the range $0 < \lambda < 1$ into the temperature solution $f_T(NTU, \tau_i, \tau_{wi}, \tau_{wo}, \lambda)$.

18.4 Defining equations and discretization

18.4.1 Ideal gas law

Mass element m [kg] of working fluid is subject to the equation of state $pV = mRT$. As pressure p, volume V and temperature T change with time t (and with crank angle $\varphi = \omega t$) dm changes accordingly. Differentiating gives $dp/p + dV/V = dm/m + dT/T$. To avoid potential problems with V approaching zero the working form multiplies through by V and transfers unknowns to the left-hand side.

$$-Vdm/m - VdT/T + Vdp/p = -dV \tag{18.7}$$

The eventual cycle description should apply to any (ideal) gas pressurized to any realistic level p_{ref} in an engine of any swept volume V_{sw}. This is handled by re-expressing in terms of normalized variables $- p/p_{ref} = \psi$, for example. Variable or 'live' volume V is normalized by V_{sw} to become $\mu = V/V_{sw}$, temperatures T by T_C, that is, $T/T_C = \tau$. The definitions of specific mass σ and mass rate $\sigma' = d\sigma/d\varphi$ are re-introduced (Equations 10.1 and 10.2).

Normalized in this way Equation 18.7 applies to both variable-volume spaces:

$$-\mu_e d\sigma_e/\sigma_e - \mu_e d\tau_e/\tau_e + \mu_e d\psi/\psi = -d\mu_e \tag{18.7a}$$

$$-\mu_c d\sigma_c/\sigma_c - \mu_c d\tau_c/\tau_c + \mu_c d\psi/\psi = -d\mu_c \tag{18.7b}$$

18.4.2 Energy equation – variable-volume spaces

In general

$$dq + dH - dW = dU \tag{18.8}$$

Increment of heat exchange dq is associated with increment $d\varphi = \omega dt$, as are increments in all other quantities.

Under present assumptions (adiabatic working spaces) dq is zero. dW is equal to pdV, dH to $c_p Tdm$, where T is the temperature of the gas element entering or leaving, and dU is $c_v d(mT) = c_v(Tdm + mdT)$, where T in this case is the temperature of the cylinder contents, assumed uniform at any instant:

$$c_p dmT|T_x - pdV = c_v(Tdm + mdT)$$

The convention T/T_x has been used to signify that in-cylinder gas bulk temperature T applies during outflow ($dm < 0$), while exchanger exit temperature T_x applies to inflow ($dm \geq 0$). Specific gas constant R is expressed in terms of isentropic index γ as $R/c_p = (\gamma - 1)/\gamma$ or $R/c_v = \gamma - 1$, as appropriate. Making the substitutions, normalizing and indicating expansion space conditions by subscript $_e$:

$$d\sigma_e(\gamma\tau_{xe}|\tau_e - \tau_e) - \sigma_e d\tau_e = (\gamma - 1)\psi d\mu_e \tag{18.9a}$$

The expression for the compression space is the mirror image:

$$d\sigma_c(\gamma\tau_{xc}|\tau_c - \tau_c) - \sigma_c d\tau_c = (\gamma - 1)\psi d\mu_c \tag{18.9b}$$

Mass M taking part in the cycle is conserved by imposing $dm_e + dm_c + dm_r = 0$. m_r is the instantaneous mass content of the regenerator. In terms of normalized variables:

$$d\sigma_e + d\sigma_c + d\sigma_r = 0 \tag{18.10}$$

18.5 Specimen implementation

18.5.1 Authentication

A matter which cannot be over-emphasized is that of *authentication*: the equations developed to this point *do not define the gas process interactions occurring within a Stirling engine.* It follows that their eventual solution will not amount to a depiction of those processes.

The solutions method must, however, process faithfully the compromized reality built into the choice of algebra – together with all the flaws. The clearest illustration will come from applying the method to the most basic possible physical situation.

The opposed piston layout is assumed. Volumes $V_e(\phi)$ and $V_c(\phi)$ vary in simple-harmonic fashion with ϕ and with constant phase difference α and with amplitude ratio $V_C/V_E = \kappa$. The two spaces connected via an annulus of diameter D [m] and radial width g [m]. The layout therefore has all the features of the coaxial (beta) 'hot air' engine, except that the regenerative annulus is stationary.

Applying the equation set gives six unknowns to be determined at each integration step, *viz* $d\sigma_e$, $d\sigma_c$, $d\sigma_r$, $d\tau_e$, $d\tau_c$ and $d\psi$. At first sight there are only five equations – 18.7a, 18.7b, 18.9a, 18.9b, and 18.10. However, the instantaneous mass content of the regenerator can be expressed in terms of gas temperature distribution $T(x)$:

$$m_r = \frac{pV_{dr}}{R} \int \frac{dx}{T(x)}$$

From Equation 18.5:

$$m_r = \frac{pV_{dr}}{RT_C} \int \frac{d\lambda}{\tau_w(x) - (\partial\tau_w/\partial\lambda)/NTU + \{\varepsilon_{\tau i} + (\partial\tau_w/\partial\lambda)/NTU\}e^{-\lambda NTU}}$$

A sleight of hand provides the extra equation. If m_r is a function of T and p, then by definition:

$$dm_r = (\partial m_r/\partial T)dT + (\partial m_r/\partial p)dp \qquad (18.11)$$

Equation 18.11 is now linear in the unknowns m_r, dT, and dp.

If the expression for m_r cannot be differentiated analytically, then the numerical counterpart, (in terms of the normalized variables to be employed here) is:

$$d\sigma_r \approx \{[\sigma_r(\tau + \Delta\tau) - \sigma_r(\tau - \Delta\tau)]/2\Delta\tau\}d\tau$$
$$+ \{[\sigma_r(\psi + \Delta\psi) - \sigma_r(\psi - \Delta\psi)]/2\Delta\psi\}d\psi \qquad (18.11a)$$

Numerical values of the perturbation quantities $\Delta\tau$ and $\Delta\psi$ are open to free choice, and are conveniently 0.05 or so of those of parent variables τ and ψ.

There are now six unknowns and six equations. However, the unknown $d\sigma_r$ is easily eliminated between Equations 18.10 and 18.11a so solution can proceed in terms of the remaining five.

18.5.2   Function form

The relationship of instantaneous pressure (or mass rate) to crank angle according to the Schmidt formulation can be expressed in function form:

$$\psi = \psi(N_T, \kappa, \alpha, \nu, \varphi) \qquad (18.12)$$

In Equation 18.12 ν is dimensionless dead space.

Application of the replacement formulation to the rudimentary gas path has introduced additional parameters:

$$\psi = \psi(N_T, \kappa, \alpha, \gamma, \nu_r, NTU, \varphi) \qquad (18.13)$$

In principle, supplying a specimen value for each permits numerical solution to proceed. Doing so generates regenerator temperature distributions which cannot possibly reflect reality: the NTU are far from constant over a cycle involving two flow reversals. It is necessary to calculate NTU from local flow conditions – that is, in terms of Reynolds number Re as solution proceeds.

18.5.3   Reynolds number in the annular gap

The relevant definition is:

$$Re = \frac{4\rho\,|u|\,r_h}{\mu} \qquad (18.14)$$

r_h of the annular gap is $\frac{1}{2}g$ [m]. The term $\rho|u|$ may be replaced by m'/A_{ff}, where m' is instantaneous mass rate, and where A_{ff} is free-flow area, equal in turn to πDg. Substituting and multiplying numerator and denominator by $\omega^2 RT_C \rho_{ref} V_{sw}$:

$$Re = \frac{2\sigma' N_{RE}}{\pi D / V_{sw}^{1/3}} \qquad (18.14a)$$

The Stanton number St is a function of Re, and NTU is defined as StL/r_h – or, in this case, as $2StL/g$. From Chapter 10 $N_{RE} = N_{SG}N_{MA}^2$. This time taking the temperature solution as the example, the function form becomes:

$$\tau(\lambda, \varphi) = \psi(N_T, \kappa, \alpha, \gamma, N_{SG}, N_{MA}, D/V_{sw}^{1/3}, g/L_r, L_r/D, \lambda, \varphi) \qquad (18.15)$$

(The presence of $D/V_{sw}^{1/3}$, g/L_r, L_r/D makes ν_r redundant.)

18.6   Integration

The coefficients of the unknowns form a 5×5 array:

$d\sigma_e$	$d\sigma_c$	$d\tau_e$	$d\tau_c$	$d\psi$	RHS
$a_{1,1}$	0	$a_{1,3}$	0	$a_{1,5}$	b_1
0	$a_{2,2}$	0	$a_{2,4}$	$a_{2,5}$	b_2
$a_{3,1}$	0	$a_{3,3}$	0	0	b_3
0	$a_{4,2}$	0	$a_{4,4}$	0	b_4
$a_{5,1}$	$a_{5,2}$	0	0	0	0

What follows is a hybrid of conventional algebra and the computer code sequence (FORTRAN) which feeds the coefficient array to a library subroutine (SIMQX) for 'solution' (Gauss–Seidel iteration).

Coefficients of gas law

– expansion space:

$$a_{1,1} = \mu_e/\sigma_e$$
$$a_{1,3} = \mu_e/\tau_e$$
$$a_{1,5} = -\mu_e/\psi$$
$$b_1 = d\mu_e + \mu_e d\mu_e d\psi - \mu_e d\sigma_e d\tau_e$$

– compression space:

$$a_{2,2} = \mu_c/\sigma_c$$
$$a_{2,4} = \mu_c/\tau_c$$
$$a_{2,5} = -\mu_c/\psi$$
$$b_2 = d\mu_c + \mu_c d\mu_c d\psi - \mu_c d\sigma_c d\tau_c$$

Mean mass rate $\underline{\sigma} = d\sigma/d\phi$ in left-right direction of flow:

$$\underline{\sigma'} = \tfrac{1}{2}(-\sigma'_e + \sigma'_c)$$
$$Re = 2.0|\underline{\sigma'}|N_{RE}/(\pi D/L_{ref})$$
$$St = 0.023Re^{-0.2}$$
$$\text{If } Re < 1042 \text{ then } St = 6.0/Re$$
$$NTU = StL_r/r_h$$

Energy equation

– expansion space:

$$a_{3,1} = -\tau_e$$

If $\underline{\sigma'} > 0$ then outflow:

$$\tau_{xe} = \tau_e$$
$$a_{3,1} = a_{3,1} + \gamma\tau_{xe}$$

else inflow:

$$a_{3,1} = -\tau e$$
$$\tau_i = \tau_c$$
$$\tau_{wi} = 1.0$$
$$\tau_{wo} = N_T$$
$$\tau_{xe} = f_T(NTU, \tau_i, \tau_{wi}, \tau_{wo}, N_T)$$
$$a_{3,1} = a_{3,1} + \gamma\tau_{xe}$$
$$a_{3,3} = -\sigma_e$$
$$b_3 = (\gamma - 1)\psi d\mu_e$$

– compression space:

$$a_{4,2} = \tau_c$$

If $\underline{\sigma'} < 0$ then outflow:

$$\tau_{xc} = \tau_c$$
$$a_{4,2} = a_{4,2} + \gamma\tau_{xc}$$

else inflow:

$$a_{4,2} = -\tau_c$$
$$\tau_i = \tau_e$$
$$\tau_{wi} = N_T$$
$$\tau_{wo} = 1.0$$
$$\tau_{xc} = f_T(NTU, \tau_i, \tau_{wi}, \tau_{wo}, 1.0)$$
$$a_{4,2} = a_{4,2} + \gamma\tau_{xc}$$
$$a_{4,4} = -\sigma_c$$
$$b_4 = (\gamma - 1)\psi d\mu_c$$

Mass inventory:

$$a_{5,1} = 1.0$$
$$a_{5,2} = 1.0$$

If $\underline{\sigma'} > 0$ then flow L − R:

$$\sigma_{r\phi} = f_m(NTU, \tau_e, N_T, 1.0)$$
$$\Delta\tau_e = 0.05\tau_e \text{ (for example)}$$
$$\sigma_{r\phi} + d\phi = f_m(NTU, \tau_e + d\tau_e, N_T, 1.0)$$
$$d\sigma_r/\Delta\tau_e = \delta_r(\sigma_{r\phi} + d\phi - \sigma_{r\phi})/\Delta\tau_e$$
$$a_{5,3} = d\sigma_{r\phi}/\Delta\tau_e$$

Otherwise flow R − L:

$$\sigma_{r\phi} = f_m(NTU, \tau_c, 1.0, N_T)$$
$$\Delta\tau_c = 0.05\tau_c \text{ (for example)}$$
$$\sigma_{r\phi} + d\phi = f_m(NTU, \tau_e + d\tau_e, 1.0, N_T)$$
$$d\sigma_r/\Delta\tau_c = \delta_r(\sigma_{r\phi} + d\phi - \sigma_{r\phi})/\Delta\tau_c$$
$$a_{5,4} = d\sigma_r/\Delta\tau_c$$
$$d\sigma_r/\Delta\psi = \delta_r/\sigma_{r\phi}$$

At each crank angle increment $\Delta\phi$ the coefficients of the unknowns are fed into the dummy arguments of SIMQX which returns solutions – the $d\sigma_e$, $d\sigma_c$, $d\tau_e$, $d\tau_c$ and $d\psi$ – corresponding to the new value of ϕ.

Because some right-hand sides (some b_i) have embodied approximate values of the unknowns the process can be iterated. Upon satisfactory convergence the new values of the dependent variables follow as:

$$\sigma_e = \sigma_e + d\sigma_e$$
$$\sigma_c = \sigma_c + d\sigma_c$$
$$\tau_e = \tau_e + d\tau_e$$
$$\tau_c = \tau_c + d\tau_c$$
$$\psi = \psi + d\psi$$

The change in mass of the regenerator $d\sigma_r$ is back-calculated:

If $\underline{\sigma'} > 0$ then

$$d\tau_r = d\tau_e$$

else

$$d\tau_r = d\tau_c$$

end if

$$d\sigma_r = (d\sigma_r/\Delta\tau)d\tau + (d\sigma_r/\Delta\psi)d\psi$$
$$\sigma_r = \psi\,\delta_r\,\sigma_{r\varphi}$$

18.7   Specimen temperature solutions

The temperature reliefs which follow have been generated for the parameter values listed in Table 18.1. All plots show the start-up and second cycles.

Chapter 20 will look at the 'real' coaxial (beta) hot-air engine, in which the regenerative effect is achieved in the radial gap between displacer and cylinder. It will be shown that

Table 18.1   Parameters values used in the construction of Figure 18.3

N_T	3.012
κ	1.0
α	90 [degrees]
γ	1.4
N_{SG}	1.404E+09
N_{MA}	1.596E–02
$(N_{RE}$	35790)
$D/V_{sw}{}^{1/3}$	1.077
$(g/L_r$	5.0E–03
L_r/D	2.0

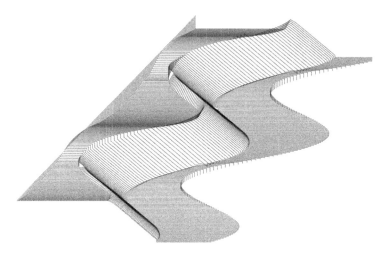

Figure 18.3a   Radial gap $g = 0.05$ mm

Reynolds number is essentially independent of g. The same is true here. In neglect of pumping losses (the choice for the present formulation) the influence of g is through through the relationship $NTU = 2StL_r/g$. All other things being equal, a change in g thus brings about an *inversely* proportional change in NTU. The effect on solutions should be an all-round increase in temperature error ε_T with increasing g.

The starting assumption of invariant, linear wall temperature gradient is tantamount to giving the regenerator infinite thermal capacity ratio. This in turn means that, for NTU infinite, it will supply and absorb the heat necessary to maintain the status quo – to cause second and successive cycles to replicate the start-up cycle. With decreasing NTU an increasing number of cycles is required for attainment of cyclic equilibrium.

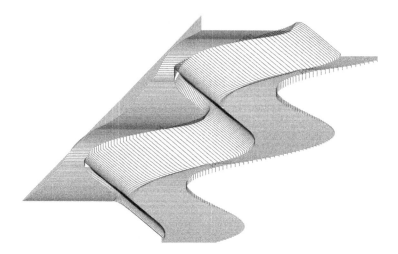

Figure 18.3b   Radial gap $g = 0.1$ mm. Marginal change relative to previous solution

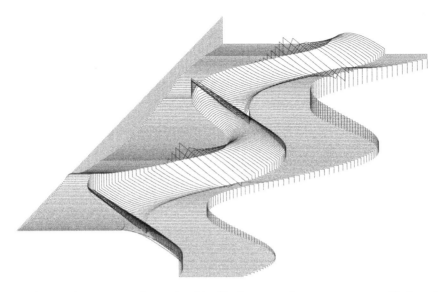

Figure 18.3c Radial gap $g = 0.5$ mm. Considerable departure of gas temperature profile from that of the wall. Temperature swing of second cycle between different limits from that of the first

For Figure 18.3a, g/L_r is set at 5.0×10^{-3} [−], or g of 0.05 mm. This unrealistically small value gives a singular peak *NTU* of 7700 at crank angle φ of 148 degrees (flow close to stationary) but values generally closer to 20. *NTU* are thus somewhat short of infinity, but the departure of the second cycle from the first is minimal. The cyclic adiabatic temperature swing shows up clearly.

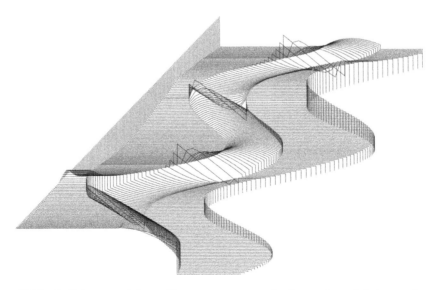

Figure 18.3d Radial gap $g = 1.0$ mm. Temperature recovery inadequate for satisfactory performance. Temperature swing in expansion space during second cycle differs little from that of compression space

Increasing g to 0.1 mm (still impracticably small, and giving peak NTU of 3850, 'mean' of 10) brings about little deterioration in terms of observable temperature error ε_T – Figure 18.3b.

Figure 18.3c is the response when g increases into the practical range at 0.5 mm. The adiabatic swing remains in evidence, but between different limits in the second cycle, which no longer resembles the first. Individual temperature profiles correspond to the specimens of Figure 18.2.

Further increase in g (Figure 18.3c and d) causes further departure of the regenerator profiles from wall temperature distribution, except at instants of flow reversal where NTU are peak, and where the plotted profiles therefore appear suspended in mid-air.

19

H₂ versus He versus air

19.1 Conventional wisdom

'Hydrogen gives superior performance to helium; helium gives superior performance to air.'[1]

At a value of 0.0821 the Beale modulus N_B of the USS H₂ charged P-40 engine falls short of the 0.129 of Philips' MP1002CA air-charged engine. What is superior about that?

If a engine operating between given temperature limits is charged first of all with air (or N₂) and then with H₂, the cycle of internal gas processes will remain nominally the same provided operating conditions (*rpm* and charge pressure) are adjusted so as to maintain unchanged the cycle distribution of Reynolds numbers *Re*. In the circumstances (engine geometry unchanged) this is achieved by holding Reynolds parameter N_{RE} unchanged:

$$N_{RE} = \frac{p_{ref}\omega V_{sw}^{2/3}}{\mu_0 R T_C}$$

Equating the two sets of operating conditions, noting that angular speed ω and *rpm* are in proportion and inserting respective numerical values of μ_0 and R:

$$\frac{(rpm \cdot p_{ref})^{air}}{(rpm \cdot p_{ref})^{H_2}} \approx 0.14 \tag{19.1}$$

According to Equation 19.1, achieving a nominally identical gas process cycle at a common value of p_{ref} (*thereby keeping computed Beale modulus N_B and internal thermal efficiency unchanged*) means operating the air-charged engine at 0.14 of the *rpm* of the same machine charged with H₂. Alternatively, at common *rpm* (if achievable) p_{ref}^{air} must be set to $0.14\, p_{ref}^{H_2}$.

---

[1] Helium cannot be taken into the comparison at this stage, being monatomic ($\gamma = 1.66$) whereas both air and H₂ are diatomic, with $\gamma = 1.4$.

Stirling Cycle Engines: Inner Workings and Design, First Edition. Allan J Organ.
© 2014 John Wiley & Sons, Ltd. Published 2014 by John Wiley & Sons, Ltd.

The *rpm* comparison at 3000 for H_2 is 420 for air. The figures have a familiar ring. On the other hand there is nothing to say:

> that either engine must operate at any particular value of N_B, or

> that an air-charged engine of 150 cm^3 swept volume must bear a physical likeness to an H_2 charged machine of the same cubic displacement: a mountain bike and a touring machine have the same number of wheels, but there the similarity ends.

19.2   Further enquiry

The physical process underlying operation is heat exchange by forced convection.

For steady, uni-directional flow of density ρ at velocity u in a duct of uniform wall temperature T_w, a local energy balance for the case of negligible kinetic energy is:

$$hp_w \cdot dx(T - T_w) = -u\rho c_p A_{ff}(dT/dx)dx \tag{19.2}$$

Noting that A_{ff}/p_w is defined as hydraulic radius r_h, and that $h/\rho|u|c_p$ is Stanton number *St*:

$$dT/dx = -St(T - T_w)/r_h \tag{19.3}$$

Equation 19.3 can be generalized somewhat by re-expressing in terms of fractional duct length $\lambda = x/L_x$:

$$\begin{aligned} dT/d\lambda &= -StL_x/r_h(T - T_w) \\ &= -NTU(T - T_w) \end{aligned} \tag{19.4}$$

Equation 19.4 states that, all things being equal, the rate at which fluid temperature T adjusts to (uniform) wall temperature T_w is proportional to number of transfer units, $NTU = StL_x/r_h$.

Now consider the accompanying pressure drop Δp over element dx of duct length:

$$\Delta p = -\tfrac{1}{2}\rho u^2 C_f dx/r_h$$

For the ideal gas, $\rho = p/RT$, and $u^2/(\gamma RT)$ is the square of Mach number *Ma*. Substituting, and again working in terms of fractional length $\lambda = x/L_x$ gives an expression for fractional pressure drop:

$$\Delta p/p = -\tfrac{1}{2}\gamma Ma^2 C_f L_x/r_h \cdot d\lambda \tag{19.5}$$

For a gas having Prandtl number $Pr \approx$ unity, Reynolds' analogy for the turbulent range has $C_f = 2St$. This enables Equation 19.5 to be written:

$$\Delta p/p = -\gamma Ma^2 NTU \cdot d\lambda \tag{19.6}$$

Re-expressed in terms of similarity parameters $Ma = N_{\mathrm{MA}}(L_x/L_{\mathrm{ref}})\gamma^{-1/2}\{\tau^{1/2}\sigma'/(\delta_x\psi)\}$. Equation 19.6 then becomes:

$$\Delta p/p = -(N_{\mathrm{MA}}L_x/L_{\mathrm{ref}})^2\{\tau\sigma'^2/(\delta_x\psi)^2\}NTU\cdot d\lambda \tag{19.7}$$

Intuitively, exchanger performance is optimal when rate of temperature adjustment (Equation 19.4) is a maximum, and associated specific pressure drop (Equation 19.7) a minimum. This suggests isolating the condition which maximizes the ratio of Equation 19.4 to Equation 19.7. (This is *not* the same thing as the misguided regenerator design criterion of Ruehlich and Quack (undated), who optimistically[2] propose the minimum of the ratio C_f/St).

For a given exchanger (e.g., expansion) under given operating conditions, the term in curly braces (Equation 19.7) is essentially independent. The $NTU\cdot d\lambda$ cancel, and the *maximum* value of the ratio Equation 19.4/Equation 19.7 is given by the *minimum* of $N_{\mathrm{MA}}L_x/L_{\mathrm{ref}}$. From the definition of N_{MA}, this requires $\omega L_x/\sqrt{(RT_C)}$ to be minimum.

If ω (or *rpm*), L_x and T_C, are fixed, the order of preference of working fluid is therefore H$_2$ ($R = 4120$ J/kgK), He ($R = 2080$ J/kgK) and air/N$_2$ ($R = 287$ J/kgK) – as decreed by Stirling lore, and consistent with the quote with which this chapter opened.

However, for any specified R and corresponding numerical value of the group $\omega L_x/\sqrt{R}$, there is a combination of ω and L_x which will yield the same numerical value when R is changed:

'For any cycle of gas processes undergone by hydrogen, there is (in principle) a gas path design yielding the identical process sequence, and the same (computed) Beale modulus N_B and the same indicated power on air.'

The opening quotation is therefore correct, and also incorrect – although it has to be said that an engine refuting it might be a challenging manufacturing proposition.

The chapter which follows will focus on the 'hot air' engine. Notwithstanding the earlier arguments, some may view this is retrograde relative to the impressive standards of efficiency and specific power achieved at NV Philips Gloeilampenfabrieken with their exclusive focus on light gases –H$_2$ and He.

19.3   So, why air?

The ultimate goal of serious Stirling engine study must surely be achievement of a range of commercially viable units.

A commercial product is not achieved by connecting a helium cylinder to an air-charged prototype via a length of flexible hose – as a glance at the external plumbing on a USS P-40 or a Solo V-160 will confirm. A proprietary Stirling engine (as opposed to the laboratory or demonstration unit) comprises at least three systems of high sophistication, complexity and cost: (1) engine per se, (2) heating provision, and (3) air pre-heater. An essential safety feature of the commercially successful unit will be micro-processor feedback between heater temperature and fuel supply rate, possibly integrated with measures for emissions handling.

---

[2] As the present treatment demonstrates, St alone does not define temperature recovery ratio, and C_f in isolation does not define pumping penalty.

Effective exhaust gas re-circulation will call for real-time monitoring. Providing H_2/He reservoir/replenishment means a four system package. Prospects in terms of cost, reliability and packaging envelope are not improved.

All conventional power units in widespread commercial use function on common, freely available working fluids: air for petrol, diesel, and gas turbine engines, water for steam engines (reciprocating and turbine). The Stirling engine starts with built-in handicap baggage of sophistication and exclusivity – materials of construction, seals, means for separating lubricant from working fluid, and so on. The additional sophistication of He or H_2 as working fluid hardly improves launch prospects.

Stirling engines in the ½–2.0 hp range (if/as/when available) will probably be of single-cylinder coaxial, or two-cylinder V-configuration. Even with H_2 as working fluid, specific power (kW/kg or hp/lb) is uncompetitive, as may be seen by applying the Beale number criterion to the reciprocating internal combustion engine. Typical Stirling value: 0.15. For the naturally-aspirated ICE the value is closer to 5 – a factor of 33 greater! In other words, this is not about competing with the ICE: The commercial Stirling engine must sell on the strength of potential for quiet, vibration-free, constant-speed, fixed-platform operation – possibly helped in some applications by its relative indifference to fuel. If this means DCHP (domestic combined heat and power), then the modest shaft speed of 1500 *rpm* becomes attractive, not merely for synchronization with the European distribution grid, but because, in the domestic application (frequently requiring wall-mounting) low noise, low vibration and reliability are paramount.

Half a century ago, Philips discovered the (undisputed) merits of H_2, He – and the dangers of air in combination with liquid hydrocarbon lubricants. They then gave up on air. This suggests that the full potential of air remains to be explored. Accounts of air-charged engines having been 'designed' in any sense are rare if not non-existent. Recent gas process modelling suggests untapped potential at useful *rpm* – say 1500–2000 for air- or N_2-charged units up to 1 kW.

20

The 'hot air' engine

20.1 In praise of arithmetic

The gas process interactions by which a Stirling-type engine converts heat to work are unde-
niably complex. Sophisticated digital arithmetic is required to gain even superficial insight.

Conversely, there is a point of view that the engine builder's priority is not insight per se,
but rather avoidance of disappointment (and/or embarrassment) when the time comes to test
his/her design. If a couple of ruthlessly basic sums can help to achieve this, then analytical
fine-points can wait.

What follows relates to *any* Stirling engine, but the principal value lies in application to the
simple hot-air configuration – for example, to the vertical, coaxial layout.

Suppose the designer is aiming at shaft power of W' watts (and that he/she has at least
checked that the combination of W', *rpm* and charge pressure p_{ref} do not violate the Beale
number criterion). If the engine eventually runs under these conditions and delivers W', the
average rate at which heat passes through the metal of the hot end into the gas is Q_E' [W].
(Q_E' will be less – almost certainly *substantially* less – than the heat rate of the combustion
system. More later.)

If the engine keeps going for long enough to attain a steady state, there will be engine
component(s) rejecting heat over a range of temperatures, but each at a steady rate. Denote
this total heat rate Q_{rej}' [W]. The majority of Q_{rej}' should occur at the heat sink provision and
therefore somewhere near ambient temperature T_C.

By definition:

$$W' = Q_E' - Q_{rej}' \tag{20.1}$$

Equation 20.1 is independent of mechanical efficiency, because heat generated by friction is
accounted for in Q_{rej}'.

Now let the internal thermal efficiency of the engine (i.e., the *actual* conversion efficiency
of the cycle of gas processes) be denoted by η_{th}:

$$\eta_{th} = W'/Q_E' = (Q_E' - Q_{rej}')/Q_E' \tag{20.2}$$

Stirling Cycle Engines: Inner Workings and Design, First Edition. Allan J Organ.
© 2014 John Wiley & Sons, Ltd. Published 2014 by John Wiley & Sons, Ltd.

Substituting Equation 20.1 into Equation 20.2 leads to $Q_{rej}' = W'(1/\eta_{th} - 1)$ or, since it is common practice to express η_{th} as a percentage:

$$Q_{rej}' = W'(100/\eta_{th}\% - 1) \qquad (20.3)$$

This is not Carnot cycle analysis. It is barely even thermodynamics – merely book-keeping.

In an era of micro-electronics, 50 W is shed-loads of power: a ruthlessly simple, reliable heat engine of that rating should be eminently marketable. Achievable internal efficiency $\eta_{th}\%$ is not known at this stage, but cannot be high: NTU in the annular gap are between 1.0 and 2.0 versus the 50–150 of a 'proper' regenerator. How about $\eta_{th}\% = 5\%$ for a start?

Substituting $\eta_{th}\% = 5.0$ into Equation 3 gives $Q_{rej}' = 19.0 \times W'$ [W] – or 950 [W]!! This is a *minimum*, since the actual value will be augmented by thermal short and by the heat equivalent of mechanical friction.

From Equation 20.1, internal heat rate $Q_E' = 1$ kW. If heating is by combustion, this grossly under-estimates external heat supply rate which, in the case of open-flame heating, could be 4–5 times Q_E'. The latter figure is consistent with the appetite for butane/propane of the Mileage Marathon engine as originally installed. (See the author's 2007 text.)

However unattractive, it is not the figure for Q_E' which should haunt the designer, but that for Q_{rej}': the former can be addressed in the short term by turning up the burner, and in the longer term by embodying an exhaust/air pre-heater. The real headache is Q_{rej}'. The incandescent light bulb is about 10% efficient, so a 100 W bulb dissipates some 90 W of heat by natural convection. The temperature of the glass can inflict a serious skin burn. It is not necessary to carry out further arithmetic to appreciate the challenge of dissipating *ten times* that amount from an engine.

The heat dissipation provision emerges as the *thermal bottleneck* of the hot air engine.

This is not (necessarily) a cue to convert from air- to water-cooling! The latter has no more than marginal impact on the cold-end thermal barrier per se. Instances are known in which a change to water cooling has led to to deterioration in overall performance!

The principal barrier is internal – between the gas (air) and solid enclosure. Water cooling can, indeed, increase the gas/wall temperature difference (by lowering effective T_C), but can also increase effective end-to-end temperature difference ($T_E - T_C$) thereby enhancing thermal short, draining heat from the expansion end.

Any solution – assuming one exists – must improve heat transfer between gas and surface of cold-end enclosure. (John Archibald has an extremely promising suggestion which will be described if permission is obtained to publicize.)

Clearly, the only realistic way forward is to increase internal efficiency. Achievable figures remain a matter of conjecture, but assume provisionally that 20% is within grasp: From Equation 20.3 $Q_{rej}' = 50 \times (100/20 - 1) = 200$ W. Recalling that the figure is a *minimum*, and that $250 \sim 300$ W are therefore more likely, it is evident that the internal efficiency target has to be raised further.

It is not necessary to have in mind a specific figure for $\eta_{th}\%$ to conclude that a 'proper' regenerator is called for. This threatens to compromise the constructional simplicity which was the original attraction of the *genre* unless a thermally efficient unit can be accommodated – a wire gauze bandage perhaps – in the necked-down diameter of the displacer.

Pending practical implementation, Figure 20.1 is the graphical embodiment of Equation 20.3. It allows quick calculation of Q_{rej}' in terms of target $\eta_{th}\%$. Noting that the $\eta_{th}\%$ scale

int_effy% [−]

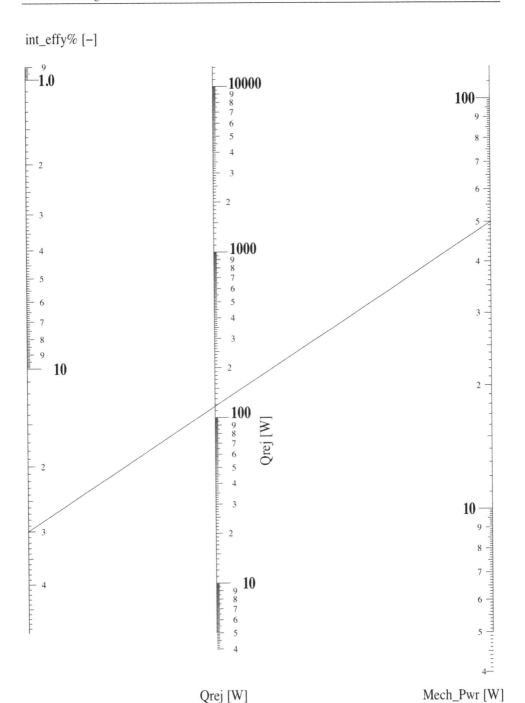

Figure 20.1 Nomogram for rapid calculation of heat rejection rate Q_{rej}' [W] in function of percentage internal thermal efficiency $\eta_{th}\%$ and mechanical power output W' [W]. The efficiency (left-hand) scale increases *downwards*

(left-hand scale) elapses *downwards* from 1% at the top to 40% at the bottom, lay a straight-edge – or draw a straight line – between chosen $\eta_{th\%}$ and target output W' [W], reading heat rejection rate Q_{rej}' [W] from the centre scale. This is a *lower bound* to the value for which the heat dissipation provision has to be designed.

The specimen line shown takes $\eta_{th}\%$ as 30%, giving $Q_{rej}' \approx 116$ W for 50W output. If the light bulb comparison is valid, then the design target which is beginning to emerge is $\eta_{th}\% > 30\%$.

20.2   Reynolds number *Re* in the annular gap

All variants of the Stirling engine exploit *forced, convective heat transfer* – a physical phenomenon carrying an inevitable penalty – *pumping power*. Best performance, whether achieved by experiment, through theory, or by both in combination, relies on striking a favourable balance between these two conflicting effects.

The parameter governing both is *Reynolds number, Re*. In this respect the hot air engine holds a surprise:[1]

Re is independent of of radial gap, *g*, between displacer and cylinder.

To confirm, start with the definition: $Re = 4\rho|u|r_h/\mu$ [−]. Hydraulic radius r_h of the annular gap is $\frac{1}{2}g$ [m], while corresponding free flow area $A_{ff} \approx \pi Dg$ [m^2] where D is displacer diameter [m].

But $\rho u A_{ff}$ is mass flow rate, $m' = dm/dt$ [kg/s]. Substituting:

$$Re \approx 2|m'|/\mu\pi D \qquad (20.4)$$

Gap *g* has disappeared – but this is not all: Re-writing m' in terms of *specific mass rate* σ' (see *Notation*) and taking the opportunity to provide temperature correction $f(\tau_{x/L})$ for the (substantial) variation in coefficient of dynamic viscosity with fractional axial location x/L_d along the gap:

$$Re = \frac{2|\sigma'|N_{RE}(L_{ref}/D)}{\pi f(\tau_{x/L})} \qquad (20.5)$$

As for the 'serious' Stirling engine the cyclic variation of σ' is essentially independent of charge pressure, p_{ref}, of *rpm*, of working gas – and of everything else. So *Re* at any given crank angle and at any given axial location within the annular gap *depends only on the parameter of operation N_{RE} and on the ratio of reference length L_{ref} (= $V_{sw}^{1/3}$) to bore D!*

But the designer requires – and has available – a means of adjusting the heat transfer intensity and its balance with flow loss. These lie within the definition of *NTU* – Number of Transfer Units, defined as the product of Stanton number *St* with ratio L/r_h. In the unique case of the hot air engine, this reduces to $NTU = 2StL_d/g$. The fact elevates the ratio L_d/g to the status of *dominant design parameter*.

---

[1]Perhaps it ought not to surprise: gap *g* doubles, particle speed |u| halves!

20.3   Contact surface temperature in annular gap

The Schmidt, the 'adiabatic' and other 'ideal' cycle analyses rely on the assumption that gas and wall temperature distributions are linear and identical between cycle temperature limits T_E and T_C.

Applied to the 'serious' Stirling engine the assumption is justifiable on the basis that design aims for that ideal. It is inappropriate to the hot air type, where the temperature gradients on inner and outer surfaces of the regenerative annulus not only differ, but are subject to substantial relative change during each cycle. This page establishes a more realistic picture, and one which serves better for first principles design.

Figure 20.2 is a simplified representation of cylinder and displacer, and indicates a notional lengthwise temperature distribution T_w for both.

The axial length of displacer is L_d; the stroke is S_d. The annular gap is formed between displacer body of length L_d and the identical length of the enclosing cylinder. However, the latter length is a fraction $L_d/(L_d + S_d)$ of total axial length $L_d + S_d$ of the enclosing cylinder. Moreover, that fraction lies at a different axial location at each instant, so that, relative to the displacer, a different enclosure temperature distribution applies.

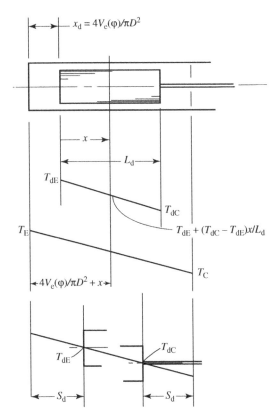

Figure 20.2   Notation for instantaneous, local effective temperature difference $\varepsilon_T = \frac{1}{2}(T_{wd} + T_{wc}) - T$ of the regenerative annulus

End temperatures of the enclosing cylinder are nominally T_E and T_C. The extremes of displacer temperature T_{dE} and T_{dC} are not known, although it is beyond doubt that $T_{dE} < T_E$ and that $T_{dC} > T_C$. It is worth proceeding despite this uncertainty, because the result will be a formulation allowing ready substitution of an improved picture as and when available.

Figure 20.2 focuses on the pair of adjacent wall temperatures T_{wd} and T_{wc} lying at axial distance x, where the datum for x is the head of the displacer. The variable axial distance x_d between this latter datum and cylinder head is acquired by inverting the expression for instantaneous expansion-space volume $V_e(\varphi) = \frac{1}{4}\pi D^2 x_d$, viz:

$$x_d = 4V_e(\varphi)/\pi D^2 \tag{20.6}$$

With ζ to denote fractional distance x/L_d:

$$T_{wd} = T_{dE} + (T_{dC} - T_{dE})\zeta \tag{20.7}$$

$$T_{wc} = T_E + (T_C - T_E)(\mu_e(\varphi)\lambda S_d/L_d + \zeta)/(S_d/L_d + 1)$$

$$\mu_e(\varphi) = V_e(\varphi)/V_{sw} \tag{20.8}$$

$$\lambda = S_p/S_d = \text{Finkelstein's kinematic volume ratio}$$

A gas element occupying the gap at fractional location ζ has temperature T, and exchanges heat with two surface elements simultaneously at different temperatures T_{wd} and T_{wc}. On the 'one-dimensional' view of traditional steady-flow heat transfer, effective temperature difference ε_T driving convective heat exchange to/from the gas element is:

$$\varepsilon_T = \frac{1}{2}(T_{wd} + T_{wc}) - T \tag{20.9}$$

Evaluating the right-hand side of Equation 20.8 for all crank-angles calls for values for T_{dE} and T_{dC}. It is assumed that T_{dE} ($< T_E$) takes the value of T_w to which the head of the displacer is exposed when at the right-hand extreme of stroke, and that T_{dC} ($> T_C$) takes the value of T_w to which the right-hand of the displacer is exposed when the displacer is at the left-hand extreme of stroke. The values are arbitrary, but allows development to proceed pending further insight.

With N_T to denote temperature parameter T_E/T_C:

$$T_{dE}/T_C = N_T + (1 - N_T)/(L_d/S_d + 1) \tag{20.10}$$

$$T_{dC}/T_C = N_T + (1 - N_T)/\{(L_d/S_d)^{-1} + 1\} \tag{20.11}$$

The algebra appears to have introduced an extra parameter L_d/S_d. This, however, is merely the product of other geometric parameters basic to the parallel bore, coaxial configuration, displacer length/diameter ratio L_d/D, bore/stroke ratio D/S_p and Finkelstein's (1960a) kinematic volume ratio λ:

$$L_d/S_d = \lambda(L_d/D)(D/S_p) \tag{20.12}$$

Figure 20.3 is a not entirely convincing attempt to portray the cyclic variation of effective gap temperature distribution $\frac{1}{2}(T_{wd} + T_{wc})$ for the numerical values of the parameters declared

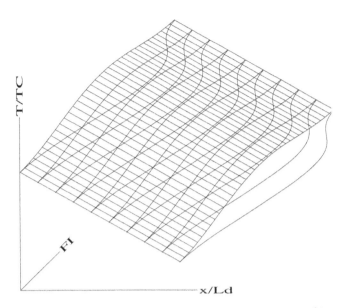

Figure 20.3 Cyclic variation of local effective surface temperature difference $\frac{1}{2}(T_{wd} + T_{wc})$ for the following parameter values: temperature ratio $N_T = 2.77$, length/diameter ratio of displacer $L_d/D = 2.5$, bore/stroke ratio $D/S_p = 1.5$

in the caption. Relative motion is displayed to scale, but is achieved by holding displacer stationary and depicting the relative motion of selected points equi-spaced axially on the cylinder (the family of lines with strong curvature). Superimposed lines having slight curvature indicate the cyclic fluctuation, referred to the 'stationary' displacer, of local effective surface temperature $\frac{1}{2}(T_{wd} + T_{wc})$.

The Schmidt, the 'isothermal' and the 'adiabatic' analysis base temperature gradient dT_w/dx on axial length of annular gap, L_d, viz, $dT_w/dx = (T_C - T_E)/L_d$ [K/m]. This overlooks the influence of substantial differences in gradient between one face of the regenerative annulus and the other, equally influential adjacent face.

Under a more realistic treatment, temperature boundary conditions of the 'ideal' cycle become a function of crank angle φ. In particular:

- The instantaneous mass inventory of the regenerative annulus is no longer a function merely of instantaneous pressure $p(\varphi)$ and of T_E and T_C.
- The working fluid entering the variable-volume spaces is no longer considered to do so at the nominal temperature limits T_E and T_C of the cycle. For values of the geometric parameters explored to date, modification of the ideal adiabatic cycle to reflect this attenuates indicated calculated cycle work by as much as 35%.

20.4   Design parameter L_d/g

1. The product of Stanton number St with ratio L/r_h is NTU. In turn the NTU value defines temperature recovery ratio in steady flow. Hydraulic radius $r_h = \frac{1}{2}g$, so $NTU = 2StL_d/g$

[−]. *St* being a function of *Re*, and *Re* being independent of gap *g* means that *L*$_d$/*g* *alone determines heat transfer in the annulus*! Moreover, under the turbulent flow conditions which must prevail, *C*$_f$ = 2*St*, and *friction also is determined by L*$_d$/*g* *alone*.

2. The optimum *g* deduced in the worked example (to follow) has a smaller value than might be suggested by experience or intuition. Benchmarks are few and far between, but Brian Thomas designed and built an engine of 5 cc nominal displacement, and it is recounted that it reached no-load speeds of 4000 *rpm*. *L*$_d$/*D* ratio was ~3.0 [−], that is, a somewhat larger value than the 2.5 of the design evolving here. Annular gap was 0.004 inch (0.1 mm), consistent with a reputation gained by Thomas for aiming for minimum gap. Measured *L*$_d$/*g* was ~656 [−]. Processing Thomas' specification via the sequence of this chapter suggests optimum *L*$_d$/*g* of ~807 [−]. This would carry a greater pumping loss than the installed value – but pumping loss was clearly not an issue at the peak brake power point of 2000 *rpm*.

20.5   Building a specification

The specific power of the un-pressurized hot air engines demonstrated to date is of the order of ½ W per cc. (The figure for the two-stroke model aero engine is a factor of 100 greater, at about 50 W/cc.) Pressurization improves the picture, but requires a closed crank-case, and introduces the challenge of sealing. All things being equal (they seldom are) a modest crank-case pressure of 4 atm gauge (5 bar abs.) promises 5 × ½ = 2½ W/cc – or 125W from an engine of *V*$_{sw}$ = 50 cc. This level of pressure looks to be within the capacity of a moulded, fibre-reinforced plastic crank-case. An engine delivering 100 We+ would extend the operating range of battery-powered mobility buggies used by the disabled, or power an entry to the Shell 'Eco-Marathon', or be a candidate for other modest duties.

This prospect defines the hot air engine for the purposes of illustrating a method of design: The focus will be on internal design rather than mechanical construction. Placing of the crank mechanism uppermost in Figure 20.4 is of no significance except for consistency with *ad hoc* heating (e.g., by a natural convection flame) for preliminary trials.

For an inspirational tour of machining and fabrication considerations, the 1993 account by Ross is unsurpassed.

Figure 20.4 might be thought of as a 'virtual' engine, since all (absolute) numerical values are left floating. Table 20.1 fleshes out the geometry somewhat (dimensionless lengths – an angle is the ratio of lengths). The choice of values for kinematic volume ratio λ and kinematic phase angle β are not 'optima' by any quantitative criterion. They are, however, consistent with low compression ratio and corresponding low pressure swing. This should maximize prospects for easy starting and for operation at intermediate temperatures.

The specification remains 'virtual' because it is independent of *rpm*, charge pressure *p*$_{ref}$ and swept volume *V*$_{sw}$. The only 'real' data are temperatures *T*$_E$ and *T*$_C$ lying behind the ratio *N*$_T$ = *T*$_E$/*T*$_C$: these are respectively 923 and 650 K (650 and 60 °C). These data suffice to kick off the two-stage design process, the first of which is pre-computation of the 'invariants' – specific mass σ, mass rate σ′ = dσ/dφ, specific pressure ψ, and pressure rate dψ/dφ of the corresponding ideal cycle.

This yields a *first approximation* to crank-angle/location histories of pressure, gas temperature, and mass fraction of any engine covered by the parameters of Table 20.1 – for example, a 100 W unit of 50 cc swept volume to operate at 1500 *rpm*.

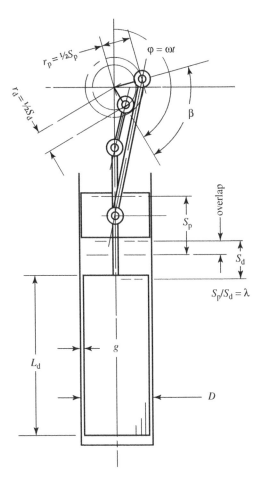

Figure 20.4 Finkelstein's (1960) notation for parallel-bore 'beta' configuration supplemented by nota-tion required by the design charts. The layout maximizes power-producing potential per net cubic displacement, while minimizing the number of parameters required for thermodynamic study – and thus for thermodynamic design

Table 20.1 Specification of a 'virtual' engine

Configuration:	coaxial, parallel-bore 'beta' (Figure 20.4)
Temperature ratio $N_T = T_E/T_C$:	2.77 [–]
Bore/stroke ratio D/S_p:	1.5 [–]
Kinematic volume ratio $\lambda \approx S_p/S_d$:	0.75 [–]
Kinematic phase angle β:	60 degree
(*equivalent thermodynamic volume ratio* κ:	~0.8 [–]
equivalent thermodynamic phase angle α:	~130 *degrees*)
Expansion space dead space ratio $\delta_e = V_{de}/V_{sw}$:	≤0.05 [–]
Compression space dead space ratio $\delta_c = V_{dc}/V_{sw}$:	≤0.05 [–]

Pressurization: via enclosed crank-case (not illustrated).

Look ahead to the eventual 'real' 100 W engine and pre-suppose the *possibility* of finding the ratios of displacer length to gap ratio L_d/g and length to diameter ratio L_d/D such that the engine performs to expectation – that is, that losses due to heat transfer deficit and to pumping are minimized. Under these circumstances the thermodynamic processes of the *real engine* differ from those of the *virtual engine* **by an amount which can be *back-calculated*.**

The two-stage process limits the number of parameters required to specify the detail of the internal gas processes. The design sequence then collapses into a manageable number of charts. The feature allows the designer to change horses in mid-stream – perhaps increasing swept volume and decreasing charge pressure – without need to resort to the computer.

20.6 Design step by step

The tentative choice of operating conditions left p_{ref} to be decided. Target value [atm.] appropriate to the provisional choice of the 50 cc unit emerges on applying the Beale number criterion. This is equivalent to checking that the combination of V_{sw}, p_{ref}, *rpm* with target power does not violate any laws of nature.

Step 1 Refer to Figure 10.2 of Chapter 10, making hard copy as required. Align a straight-edge with 1500 *rpm* on the right-hand scale and 100 W on the power (*Pwr*) scale. Mark the point of intersection with the un-graduated scale. Align the straight-edge with this latter point and with 50 cc on the V_{sw} scale. Note where the straight-edge cuts the p_{ref} scale. This latter value (5.33) is the charge pressure in atm. giving *potential* to deliver 100 W at 1500 *rpm when gas path geometry has been optimized.*

*If this is the first use of a four-scale alignment nomogram, check the result by using the Beale equation in a **consistent system of units**, as below:*

$$\frac{Pwr \text{ [W]}}{p_{ref} \text{ [Pa]} \, V_{sw} \text{ [m}^3\text{]} \, rpm/60} < 0.15$$

Step 2 Calculate reference length $L_{ref} = V_{sw}^{1/3}$. The result must be in metres [m], so check that you have arrived at the value $(50 \times 10^{-6})^{1/3} = 0.03684$ [m]. Convert *rpm* to angular speed ω [rad/s] using $\omega = rpm \times 2\pi/60$ and verify $\omega = 157$ rad/s. Little is lost by using the handy approximation $\omega \approx rpm/10$, which suggests $\omega = 150$ rad/s.

Step 3 *Either*: Connect 1500 on the left-hand scale of Figure 10.3 with 50 cc on the right-hand scale and read the value of speed parameter N_{MA} from the right-hand (air) side of the centre scale;

or: recall that, for air, gas constant $R = 287$ J/kgK, take heat sink temperature to be 333 K (60 °C) and apply hand calculator or slide-rule to formula $N_{MA} = \omega L_{ref}/\sqrt{RT_C} = \omega V_{sw}^{1/3}/\sqrt{RT_C}$. Either way, confirm that $N_{MA} = 0.01871$ [–].

Step 4 *Either*: go to Figure 10.4 and read Stirling parameter N_{SG} from the left-hand (air) side of the centre scale in terms of p_{ref} and *rpm*;

or: use $\mu_0 = 0.017 \times 10^{-3}$ and apply hand calculator or slide-rule to formula $N_{SG} = p_{ref}/\omega\mu_0$. In either case, verify that $N_{SG} = 1.99 \times 10^8$ [–].

Step 5 *Either*: use Figure 10.5 to read Reynolds parameter N_{RE} in terms of N_{MA} and N_{SG};
   or: apply hand calculator or slide-rule to $N_{RE} = N_{SG}N_{MA}{}^2$. In either case, confirm $N_{RE} = 69.9 \times 10^3$.

20.7   Gas path dimensions

The Specification set the value 1.5 [–] for bore/stroke ratio D/S_p. 50 cc is the provisional choice for V_{sw}. The combination determines bore D and stroke S_p separately.

Noting that the chart of Figure 20.5 is in convenience units (mm, cc), locate the specification values of V_{sw} and D/S_p on outer scales, join with a straight line, read bore D [mm] and check that, for the present example, $D = 45.7$ mm.

Using hand-calculator or slide-rule acquire stroke $S_p = D/(D/S_p)$ [mm] $= D/1.5$ [mm] $= 30.46$ mm.

Choose a tentative value for length/diameter ratio of displacer, L_d/D [–]. Values between 2.5 and 3.0 are promising. Recall the numerical value of speed parameter N_{MA} and form the ratio L_d/D. For the present case verify $N_{MA}/(L_d/D) = 0.01871/2.5 = 0.00748$ [–].

Proceed to Figure 20.6. Recall the value of Reynolds parameter N_{RE} from earlier arithmetic. With a straight line join the value of $N_{MA}/(L_d/D)$ on the left-hand scale to that of N_{RE} on the right, and read off optimum length/gap ratio L_d/g [–] – 822.0

This is the value which, *on the basis of the standard $N_{st}Pr^{2/3}$ – Re and C_f-Re correlations for turbulent, steady flow*, minimizes the sum of cycle losses – heat transfer deficit and pumping work.

Proceed to Figure 20.7 and read loss Z_q due to heat transfer deficit, verifying $Z_q = 0.1135$ [–]. Values are normalized to the same base as ideal cycle work Z, so the penalty may be subtracted directly.

Proceed to Figure 20.8 and read loss Z_f due to pumping and verify $Z_f = 0.0462$ [–]. As in the case of Z_q, values are normalized to the same base as ideal cycle work, Z, so the penalty may be subtracted directly.

Dead space δ of the annulus as parameter. δ is evidently a function of gap g [m], defined as V_{dgap}/V_{sw}. In terms of the (dimensionless) design parameters, numerical values for which are now available:

$$\begin{aligned} \delta &= \pi D L_{dg}/V_{sw} \\ &= \pi\{(D/L_{ref})^3(L_d/D)^2(L_d/g)^{-1}\} \end{aligned} \tag{20.13}$$

From Figure 20.9, ideal Z corresponding to this value of δ is 0.32. Net, specific cycle work Z_{net} follows on subtracting losses as $Z - Z_q - Z_f$. For the L_d/D value of 2.5 chosen at random earlier $Z_{net} = 0.32 - 0.1135 - 0.0462 = 0.1603$ [–]. The corresponding power is $0.1603 \times p_{ref}$ [Pa] $\times V_{sw}$ [m^3] $\times rpm/60 = 106.8$ W.

The numerical result exceeds the initial 100 W target by some 7%, suggesting a satisfactory outcome. The apparent success of the method is subject to a reservation applying to all design carried out from first principles.

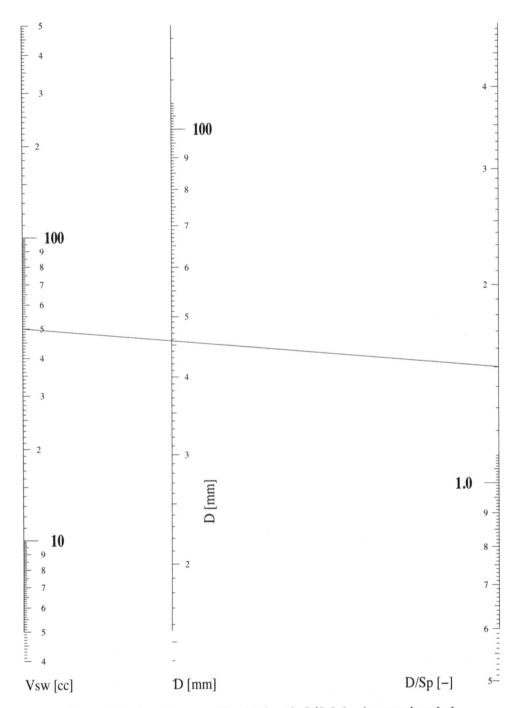

Figure 20.5 Bore D in terms of bore/stroke ratio D/S_p [−] and swept volume [cc]

NMA/(Ld/D) [−]

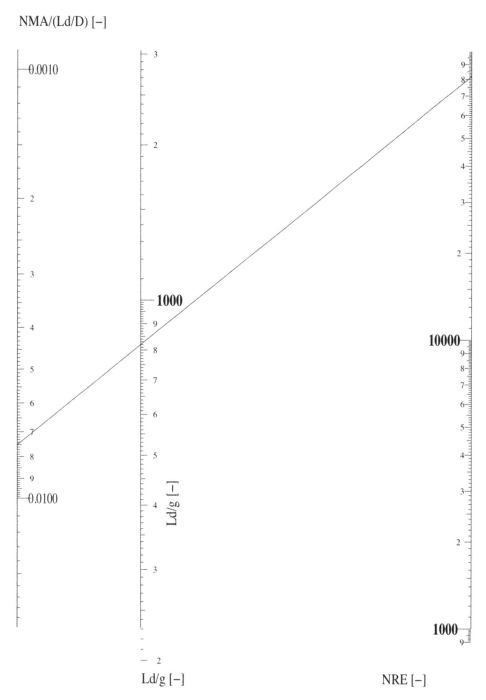

Figure 20.6 Join $N_{MA}/(L_d/D)$ of left-hand scale to N_{RE} of right-hand scale and read optimum L_d/g from centre scale

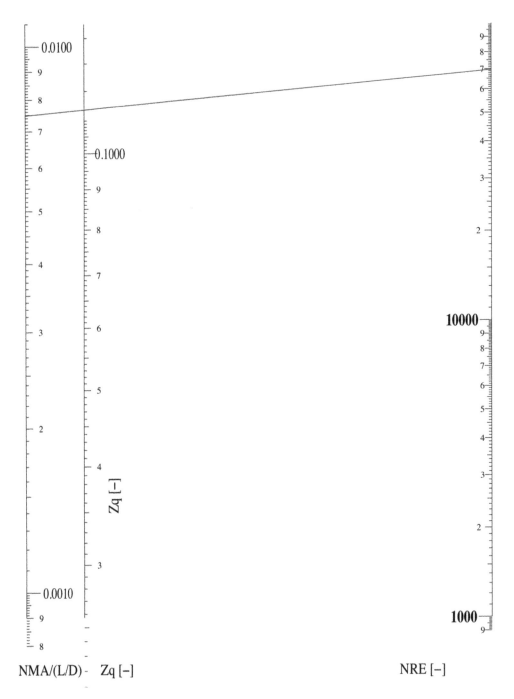

Figure 20.7   Loss due to heat transfer deficit Z_q in function of $N_{MA}/(L_d/D)$ and N_{RE}

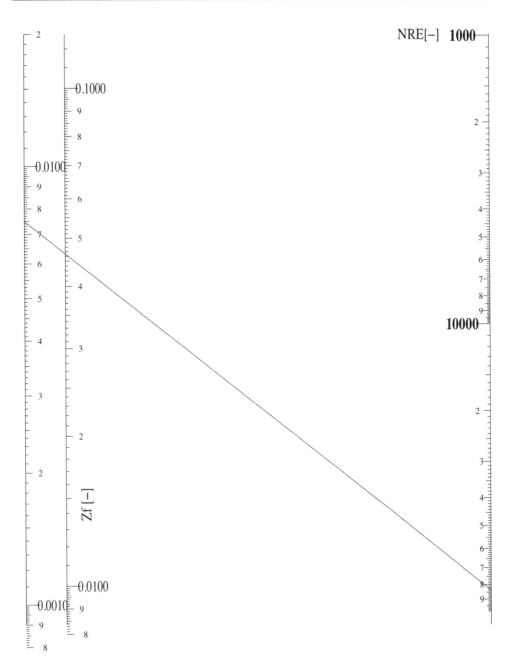

NMA/(L/D) Zf[−]

Figure 20.8 Loss due to flow friction deficit Z_f in function of $N_{MA}/(L_d/D)$ and N_{RE}

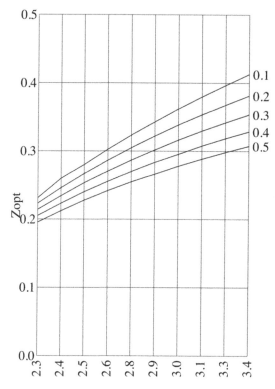

Figure 20.9   Ideal cycle work for optimum L_d/D and L_d/g as a function of temperature ratio N_T. The parameter is dimensionless dead space δ

20.8   Caveat

The value – indeed, the validity – of the method depends essentially on the relevance of steady-flow heat transfer and flow friction correlations to conditions in the regenerative annulus.

A moment's thought leads to the conclusion that the relevance, if any, is at best tenuous. So where does this leave the 100 W design? – indeed, what are the credentials of the chart-based approach?

First of all, no worse off than conventional simulation relying on the very same correlations. But then, no better off either – except that the *skeleton* per se *of the procedure* does not depend on any particular source of data. If it has attracted any appeal versus the competition (simulation, scaling, etc.) then the chances of eventually seeing an improved data resource, and thus of eventually gaining accreditation, may have improved slightly.

21

Ultimate Lagrange formulation?

21.1 Why a new formulation?

Heat transfer data presently available are irrelevant to conditions in the 'hot air' engine, whose mechanical simplicity might otherwise reward detailed analysis. A resource which, in principle, has no limitations is the *kinetic theory of gases*. Literal implementation would involve monitoring the behaviour of some 10^{25} individual molecules, but the algorithms relevant to the eventual task can be set up and exercised on a tiny fraction of that number. The exercise is a way to estimating the minimum number of molecules required to give a picture of macroscopic gas behaviour – information which might otherwise have to await development of further generations of computer.

21.2 Context

The principle of the closed-cycle regenerative engine can be demonstrated in a simpler mechanical embodiment than is possible with other reciprocating cycles. The result is the 'hot-air' engine already illustrated at Figure 20.4. This can be realized in as few as three moving parts. It is when attempts are made to extract competitive specific power that the Stirling engine becomes orders of magnitude more complex and expensive.

A power range not well served[1] by the IC engine is that between 25 and 100 W. The former figure would power a lap-top computer. The latter the portable oxygen concentrators used in medical emergency, and currently restricted to one hour's operation per 1 kg electric battery.

It is possible that, with appropriate thermodynamic design, the use of modern materials, and by resorting to pressurization, this power range would come within the scope of the 'hot-air' type. Multi-tube heat exchangers do not feature. A regenerative effect is achieved as flow is shuttled via the narrow annulus between cylinder and displacer.

The challenge of predicting performance potential involves defining stroke ratio $\lambda = S_p/S_d$, kinematic phase angle β, axial length L_d, and radial gap g which, in combination, give the

---

[1]The model aircraft engine appears to hold a monopoly – until account is taken of high shaft *rpm*, noise, exhaust, fumes, and hit and miss starting.

most favourable balance between the benefits of heat transfered over a cycle and the penalty of accompanying pumping work. The context is:

Unsteady flow between surfaces in relative motion.

Temperature distribution at inner (moving) bounding surface different from that of opposite-facing stationary surface.

Cyclic pressure out of phase with both other cyclic events – fluid flow and displacer motion.

It is ironic that conventional correlations ($StPr^{2/3}$ vs Re and Cf vs Re for steady, fully established, uni-directional flow) are even less appropriate to this most basic of configurations than to the highly sophisticated variant.

A possible recourse is to a proprietary CFD package. Developers who have responded to the author's enquiries have expressed doubt as to suitability[2] for the purpose. Regardless of proprietary assurances, application of 'black box' code without access to the underlying algorithms raises misgivings.

When all else fails one is entitled to think the unthinkable: the ultimate gas process simulation tracks the events of *individual molecules*. A hot-air engine of 50 cc nominal displacement might enclose a volume of 100 cc. At 5 atm. and with contents at standard temperature Avagadro's hypothesis suggests 1.344×10^{22} molecules.

Settling for insights on offer from a representative two-dimensional 'slice' through the gas – perhaps a few microns thick – reduces the number by 3–4 orders, but still raises the prospect of four arrays, each $10^{18} \times 10^{18}$ for dealing simultaneously with individual values of u and v at each x and y coordinate of molecule location.

Replacing the simultaneous formulation by an explicit serial solution would reduce storage requirement to a 'mere' 4×10^{18} for the number of molecules envisaged to this point. The development of such an explicit algorithm is inherent to this account.

Further economies are possible. To avoid working through a tedious list these are introduced in context as the solution strategy is presented. Specimen illustrations precede the algebra.

21.3   Choice of display

The insight on offer can be fully appreciated only by seeing the on-screen animated display which accompanies the integration process. There is a choice between two modes of display: (a) 'streak' traces of the path of each molecule, refreshed at each integration step and (b) dots to represent successive locations – again refreshed once per time step. In the form of screen-shot stills neither does justice to the animation: Option (a) is problematic because a relatively small number of streaks gives an image too dense to interpret. Option (b) allows display of an order of magnitude greater number of molecules, but is unhelpfully static. Neither succeeds in depicting the bulk motion evident from the animation.

The illustrations which follow depict the right-hand symmetrical half of the 'engine'. The un-representatively wide annular gap dimension g permits integration time step Δt to be set at a large value, speeding development of the solution algorithm.

Figure 21.1 is a specimen 'streak' frame. Some 4320 molecules have been tracked through 5¼ revolutions of the crankshaft from start-up to 5000 *rpm*. It might be anticipated that

---

[2]The majority of successful applications publicized are for unsteady but overwhelmingly uni-directional flow – although this may not be a consideration.

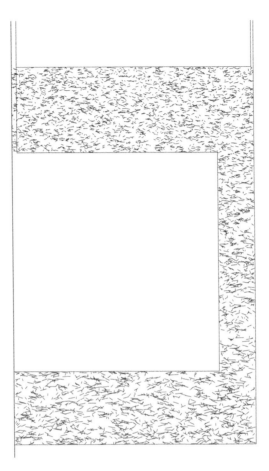

Figure 21.1 Right-hand symmetrical half of section through 'engine' of Figure 20.4, showing streak traces of each of 4320 molecules at $5\frac{1}{4}$ crankshaft rotations from instantaneous start-up

the image would be darker in the compression (upper) space where the fluid is denser. The expected pictorial effect is offset by the fact that streak elements are longer in the expansion space where molecular speed is higher.

For the numerical model to justify refinement towards serviceablity it must be capable of depicting one or more 'obvious' flow phenomena.

Mean molecular velocity for N_2 calculated from Equation 21.3 (below) at room temperature is 476 m/s. Un-surprisingly, macroscopic acoustic speed is of similar order at 353 m/s. The virtual engine can be 'overdriven' at *rpm* such that peak piston speed ωr is comparable to acoustic speed, thereby guaranteeing that the gas in the corresponding 'real' engine – supposing the latter could achieve those *rpm* – would exhibit the rarefaction waves and shocks which characterize compressible behaviour.

Start-up and acceleration to full-speed are instantaneous in the 'virtual' engine. For the 'still' of Figure 21.2 the number of molecules has been increased to 3.8×10^4 and integration frozen at $5\frac{1}{4}$ revolutions from start-up. In the animation a cyclic pattern of 'swarming' is apparent.

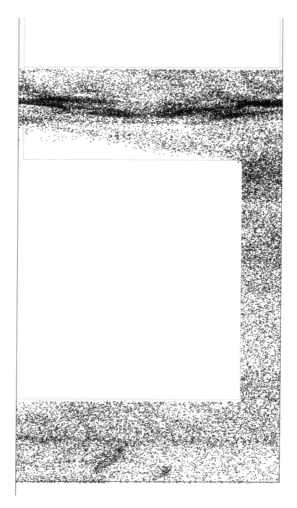

Figure 21.2 'Cloud' of 3.8×10^4 molecules at $5\frac{1}{4}$ crankshaft rotations from instantaneous start-up to 300 000 *rpm*, causing two-dimensional shock waves and rarefactions. Higher *rpm* still are required to cause similar phenomena in the expansion space, where acoustic speed is greater by a factor of $\sqrt{3}$

At the point of the cycle illustrated, displacer is at peak downwards speed. The inertia of the molecule 'cloud' entering the compression (upper) space from the annular gap causes a rarefaction above the upper horizontal face of the displacer. Heavy concentrations in the 'cloud' correspond to the opposite – elevated density and pressure.

21.4   Assumptions

Molecules are spheres in perfectly elastic collision.

Integration time step Δt can be chosen sufficiently short that there is only one collision (molecule-to-molecule, or molecule-to-enclosure) per Δt.

The method does not cope with simultaneous impact between 3 bodies.

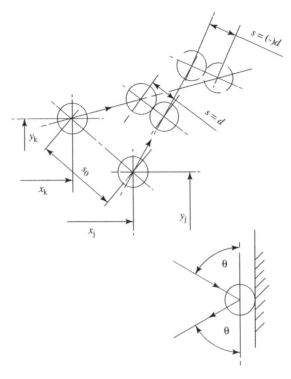

Figure 21.3 Upper: notation for collision prediction sequence. Lower: basic case of wall collision – wall stationary

The upper diagram of Figure 21.3 represents two molecules, j and k, on course to potential collision. At time $t = t_0$ particle j is at x_j, y_j and travelling with constant velocity components u_j, v_j, while k is located at x_k, y_k and travelling at u_k, v_k. At time t_0 the origins are separated by (scalar) distance s_0:

$$s_0 = \sqrt{\{(x_k - x_j)^2 + (y_k - y_j)^2\}}$$

After a small time increment dt separation s is given by:

$$s = \sqrt{\{(x_k + u_k dt - x_j - u_j dt)^2 + (y_k + v_k dt - y_j - v_j dt)^2\}}$$

If there is a value of dt ($< \Delta t$) for which separation distance s reduces to molecular diameter d, a collision occurs. Equating s to d in terms of coefficients of dt:

$$dt^2\{(u_k - u_j)^2 + (v_k - v_j)^2\} + 2dt\{(x_k - x_j)(u_k - u_j) + (y_k - y_j)(v_k - v_j)\} + s_0^2 - d^2 = 0$$

This is a quadratic equation dt of the form $a dt^2 + b dt + c = 0$ having the usual solutions (involving the square root):

$$a = (u_k - u_j)^2 + (v_k - v_j)^2$$

$$b = 2\{(x_k - x_j)(u_k - u_j) + (y_k - y_j)(v_k - v_j)\}$$

$$c = s_0^2 - d^2$$

For pairs of molecules for which the argument of the root is positive, there are two real solutions, one or both of which may be positive. If both are positive, then it is the smaller which is relevant, since the larger relates to the (virtual) instant of *separation* where distance s passes through the value d_m for a second time.

Under assumptions to this point, the algebra of wall collisions (lower diagram of Figure 21.3) is elementary.

21.5 Outline computational strategy

An integration step starts by setting moving solid boundaries (piston face and displacer outline) to respective current locations.

Molecules, identified by index j are scanned in sequence ($j = 1, n_{mol}$). Each is first examined as to whether current location x_j, y_j and velocity components u_j, v_j lead to contact with a wall. If so, point of contact and amount dt remaining out of time interval Δt are calculated, and post-contact location x_j', y_j' determined. By hypothesis this particle is not also in collision with another individual molecule during dt, so can be re-positioned.

The process repeats until a molecule-to-molecule collision with k is detected, whereupon the collision algebra (below) is applied to give u_j', v_j' and thence x_j', y_j'. By hypothesis, molecules j and k are not candidates for further collision until the next time step, and are re-positioned in anticipation.

The scan needs to cover every possible pair of molecules j and k – but there is scope for economy: suppose the first molecule of a scan, molecule $j = 1$, has been inspected for possible contact with all molecules k up to $k = 1500$ and a collision eventually registered. Molecule 1 is not allowed further collisions, so the scan recommences – but not at $j = 2$, because the potential collision $j = 1$, $k = 2$ has already been tested, but with molecule $j = 2$. Each successive scan in the sequence of n_{mol} scans per Δt is thus shorter by one than its predecessor in the sequence. Further economy is possible, as explained after an account of collision mechanics.

21.6 Collision mechanics

Figure 21.4 shows molecule j at the instant of collision with molecule k. Under present assumptions respective components of momentum in the tangential direction are unchanged by the impact. Components of momentum along a line through the origins follow the law for collinear impact:

$$u_j' = -u_k; u_k' = -u_j \tag{21.1}$$

The prime ($'$) indicates values after impact.

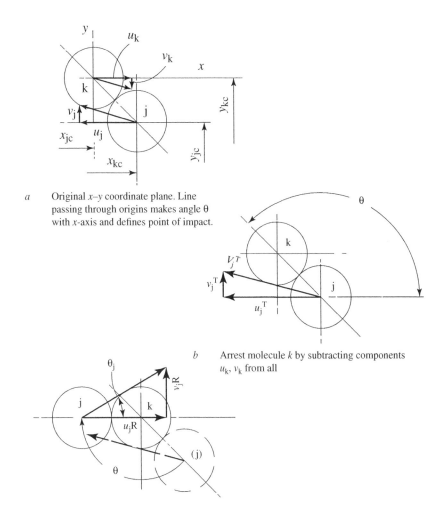

a   Original x–y coordinate plane. Line passing through origins makes angle θ with x-axis and defines point of impact.

b   Arrest molecule k by subtracting components u_k, v_k from all

c   Rotate molecule j about axis of k through angle θ

Figure 21.4   Notation for collision algorithm (*Continued overleaf*)

A way of implementing these principles is first to arrest one of the particles. In Figure 21.4 molecule k is arrested by subtracting u_k from all x components of velocity and v_k from all y-components. The step is equivalent to transformation to a new reference frame moving with components $-u_k$ and $-v_k$.

A line through both origins defines the point of contact and the directions of normal and tangential components of momentum exchange. With subscript c to denote coordinates at the instant of impact:

$$\theta = \operatorname{atan}\{(y_{kc} - y_{jc})/(x_{kc} - x_{jc})\} \tag{21.2}$$

(The numerical value of the argument of Equation 21.2 – even when properly signed – is ambiguous. Satisfactory implementation calls for the coding of a function which inspects

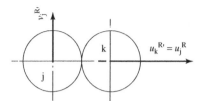

d   Apply equation for co-linear impact

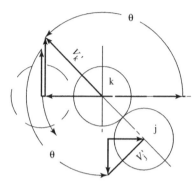

e   Rotate coordinate system through angle
        minus θ to earlier transformed plane

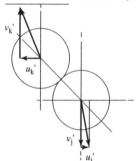

f   Add original velocity components u_k, v_k to values $u^T_j{'}$, $v^T_j{'}$, $u^T_k{'}$ and $v^T_k{'}$ after
        back-rotation to give post-impact components $u_j{'}$, $v_j{'}$, $u_k{'}$ and $v_k{'}$ in original
        plane

Figure 21.4   (*Continued*)

*numerator and denominator of the argument for the quadrant in which the angle lies. Crucial
to correct computation is choice of appropriate inequality checks* >, ≥, *etc.*)

Superscript ^{T} distinguishes values in the new reference frame:

$$u^T_j = u_j - u_k; \qquad v^T_j = u_j - u_k$$

The velocity of particle *j* in the new frame is defined in terms of speed (modulus) V_j and angle:

$$|V_j| = \sqrt{\{u^T_j{}^2 + v^T_j{}^2\}}$$

$$\theta_j = \mathrm{atan}(v_j/u_j)$$

With a view to applying Equation 24.1 the transformed coordinate system is rotated about the origin of molecule k through angle θ. The rotation has no effect on modulus $|v_j|$.

Using superscript R to denote components rotated to the new system:

$$u^R_j = |V_j| \cos(\theta_j - \theta); \qquad v^R_j = |V_j| \sin(\theta_j - \theta)$$

Re-introducing prime $'$ to denote values after impact:

$$u^R_j{}' = 0; \qquad u^R_k{}' = u^R_j;$$
$$v^R_j{}' = v^R_j; \qquad v^R_k{}' = 0$$

Post-collision moduli are unchanged by rotation back through the original contact angle θ:

$$|V^T_j{}'| = \sqrt{\{0^2 + v^R_j{}'^2\}} = v^R_j{}'$$
$$|V^T_k{}'| = \sqrt{\{u^R_k{}'^2 + 0^2\}} = u^R_k{}'$$

Respective angles required for re-orientation to the earlier transformed plane are:

$$\theta^T_j{}' = \operatorname{atan}(v^R_j{}'/u^R_j{}')$$
$$\theta^T_k{}' = \operatorname{atan}(v^R_k{}'/u^R_k{}')$$

Applying the rotations involves use of the original θ:

$$u^T_j{}' = |V^T_j| \cos(\theta^T_j{}' + \theta); \qquad v^T_j{}' = |V^T_j| \sin(\theta^T_j{}' + \theta)$$
$$u^T_k{}' = |V^T_k| \cos(\theta^T_k{}' + \theta); \qquad v^T_k{}' = |V^T_k| \sin(\theta^T_k{}' + \theta)$$

Post-impact velocity components are acquired by reversing the original transformation, that is, by adding pre-impact values of u_k and v_k:

$$u_j' = u^T_j{}' + u_k; \qquad v_j' = v^T_j{}' + v_k$$
$$u_k' = u^T_k{}' + u_k; \qquad v_k' = v^T_k{}' + v_k$$

The integration time increment Δt during which collision occurs is reckoned from t_0. Denoting time of collision by t_c, molecules j and k move apart during the balance of Δt for length of time time $d\Delta t$, where $d\Delta t = \Delta t - (t_c - t_0)$. The new locations at the end of the integration step are thus:

$$x_j = x_c + u_j' d\Delta t; \qquad y_j = y_c + v_j' d\Delta t$$
$$x_k = x_c + u_k' d\Delta t; \qquad y_k = y_c + v_k' d\Delta t$$

If a framework can be set up for economically tracking, storing and retrieving velocity and location, then individual event descriptions can be dummies, to be embellished later *ad lib*. In this spirit, Figure 21.3 illustrated a collision with a containing surface: the molecule is provisionally assumed to rebound with the x-component of velocity reversed and with the y-component un-altered – except when the surface is in motion (horizontally at u_w or vertically at v_w), in which case the appropriate velocity component is increased by u_w or v_w.

21.7   Boundary and initial conditions

Proper functioning of the numerical algorithm should be independent of number of molecules. It can therefore be explored with a small number (some hundreds) of molecules. These are initially laid out on a rectangular lattice to a density inversely proportional to absolute temperature of the immediately enclosing space – as for the 'isothermal' assumption of the conventional, macroscopic 'ideal' cycle.

Initial particle speed is set according to absolute temperature using the Maxwell–Boltzmann distribution law:

$$|\underline{V}| = \sqrt{\{8\underline{V}^2/3\pi\}}$$
$$\underline{V}^2 = 3kT/m$$

(21.3)

in which:

$$k = \text{Boltzmann's constant } (\approx 1.38 \times 10^{-23} \text{ J/K})$$

$$T = \text{absolute temperature [K]}$$

$$m = \text{mass of molecule}$$

For each molecule a random number generator allocates initial x-velocity component u such that $\sqrt{(u^2 + v^2)} = |\underline{V}|$, where v is the corresponding initial y-component. The latter is evaluated as $v = \sqrt{(\underline{V}^2 - u^2)}$. Individual molecules thus head off at speed $|\underline{V}|$ according to starting location, but in random directions.

21.8   Further computational economies

The scan is the dominant computational overhead – and offers the greatest scope for economies.

Assuming no intermediate collisions, the maximum relative distance S_{rmax} which can be covered by molecules j and k during m consecutive unit time steps Δt is $S_{rmax} = m\Delta t(|V_j| + |V_k|)$. A 'screening' scan, restricted to one per m time steps, can eliminate all molecules with which molecule j cannot possibly collide before a subsequent screening. The index of any molecule for which potential collisions are *not* eliminated is added in ascending order to a subset of indices. Some such sub-sets may contain the index for only a single potential collision. Over the subsequent m integration steps both the total number of scans and also the number of scans per molecule are both dramatically reduced.

Impact between molecules which, at the start of the screening scan are S_{rmax} apart, is possible only if they are already aligned on collision course. The statistical likelihood[3] is small, and a

---

[3]With statistics already creeping in there may be some attraction to basing this sub-algorithm on mean free path rather than on individual speeds.

way of capitalizing is to implement the screening scan for interval $m\Delta t$ – but then to continue the regular scans over a number of time steps 20% (say) greater than m.

'Collision diameter': As the thickness of the imaginary slice taken through the three-dimensional gas to give the two-dimensional picture is reduced so is the number of molecules to be tracked. A reasons for not making the number arbitrarily small is that the number of collisions of the resulting 2-D picture becomes unrepresentative. A counter-measure being explored is to calculate collision events (only) in terms of a *collision diameter* $d_c > d$ such that mean free path for the 2-D field is numerically equal to that of the 3-D case.

Scaling: The 50 cc concept of Chapter 20 was just that – a 'virtual' engine. The solution strategy can be explored without compromise by working with an engine of 5 cc displacement – or 0.5 cc – or 0.05 cc and subsequently converting to full-size by formal scaling. Storage requirement accordingly reduces by further orders of magnitude.

Processing power: CPU time in generating Figure 21.2 was 3.59 sec per revolution of the crank-shaft using an 8-year old PC with Windows XP, Pentium P 845 processor and 256 MB memory. Euler-based, conventional 'third-order' simulation code recently announced by García-Granados (2013) allegedly takes half an hour to reach cyclic equilibrium.

On this basis, and with the benefits of coding for parallel processing on dual-core hardware yet to be explored, the gas kinetic approach is under no disadvantage relative to conventional implementations.

21.9 'Ultimate Lagrange'?

It is an inherent feature of the formulation that not a single molecule 'leaks' or is lost during the integration step. The situation compares favourably against the net effect of interpolation, linearization, round-off, truncation, and so on which afflicts implementation of the traditional, macroscopic counterpart.

Appendix 1

The reciprocating Carnot cycle

In Figure A.1.1 a cylinder contains a fixed mass of gas enclosed by a piston. Ideal gas behaviour is assumed, but the assumption is not essential. The cylinder and piston are perfect thermal insulators. Three interchangeable cylinder heads are provided, one a perfect insulator, one permanently hot at absolute temperature T_E, and one permanently cold at T_C. The latter two are of unlimited thermal capacity and of infinite thermal conductivity. Switching between heads can be carried out without loss of gas or thermal leak.

All cylinder heads allow the same volumetric clearance V_3 at piston inner dead centre, and this determines compression ratio $r_v = V_1/V_3$.

Under Carnot's assumptions the means of piston actuation is immaterial. On the other hand, achieving Carnot's objective of a closed cycle requires switching cylinder heads at volumes V_2 and V_4 pre-set in terms of compression ratio r_v and temperature ratio $N_T = T_E/T_C$. Carnot omits this precaution, and his four-volume event sequence does not necessarily close to a cycle.

Rectifying Carnot's oversight is facilitated by reckoning instantaneous volume V in terms of the angular position of a crank. Providing for a degree of kinematic flexibility, as in the figure, makes no difference to the ideal cycle (isotherms and adiabats) but has a noticeable effect when the cycle is modified for the effects of finite (as opposed to infinite) heat rates.

With the above in mind, a cycle starts with the piston at outer dead-centre and with heat sink at T_C in position:

1–2 Compression with heat rejection uniformly at T_C – to volume V_2 pre-calculated so that subsequent adiabatic compression will raise T to precisely T_E when V reaches V_3

2–3 Insulating cylinder head replaces heat sink. Adiabatic compression to inner dead-centre ($V = V_3$)

3–4 Insulating cylinder head replaced by heat source at T_E. Expansion at uniform T_E to volume V_4 pre-calculated so that subsequent adiabatic expansion will lower T to precisely T_C when V reaches V_1

4–1 Insulating cylinder head replaces heat source. Adiabatic expansion to outer dead-centre ($V = V_1$).

Stirling Cycle Engines: Inner Workings and Design, First Edition. Allan J Organ.
© 2014 John Wiley & Sons, Ltd. Published 2014 by John Wiley & Sons, Ltd.

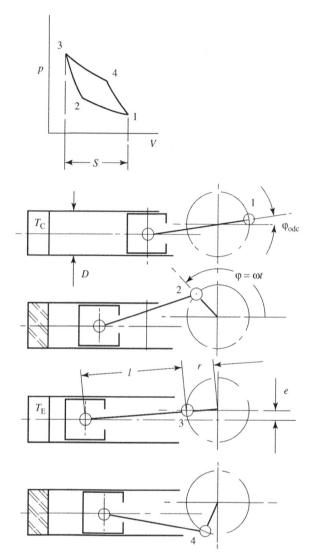

Figure A.1.1 Sequence of four process paths forming the ideal, reciprocating Carnot cycle: 1–2 isothermal compression at temperature T_C; 2–3 adiabatic compression; 3–4 isothermal expansion from and at T_E; 4–1 adiabatic expansion to starting conditions. (The sequence described by Carnot starts with isothermal expansion from and at T_E)

Appendix 2

Determination of V_2 and V_4 – polytropic processes

For the polytropic process 1–2 of Figure A.1.1:

$$T_2 V_2^{n-1} \approx T_1 V_1^{n-1}$$

For adiabatic phase 2–3:

$$T_3 V_3^{\gamma-1} \approx T_2 V_2^{\gamma-1}$$

Dividing the latter equation by the former:

$$V_2^{\gamma-1}/V_2^{n-1} = \left(T_3/T_1\right)\left(V_3^{\gamma-1}/V_1^{n-1}\right)$$

From the definition of compression ratio $V_3 = V_1/r_v$, and noting that $T_3/T_1 = N_T$:

$$V_2^{\gamma-n} = (T_3/T_1)(V_3^{\gamma-1}/V_1^{n-1})$$

$$= N_T \frac{(V_1/r_v)^{\gamma-1}}{V_1^{n-1}} \tag{A.2.1}$$

The counterpart expression for V_4 which marks the end of the polytropic expansion phase is:

$$V_4^{\gamma-n} = 1/N_T \frac{V_1^{\gamma-1}}{(V_1/r_v)^{n-1}} \tag{A.2.2}$$

Stirling Cycle Engines: Inner Workings and Design, First Edition. Allan J Organ.
© 2014 John Wiley & Sons, Ltd. Published 2014 by John Wiley & Sons, Ltd.

Appendix 3

Design charts

A.3.1 Raison d'être

It may be necessary to be an engineering designer to appreciate charts and graphs as media for interacting with one's working formulae.

The concurrent digital and analogue display of a properly-constructed chart conveys, at a single glance, the value of the underlying function or mathematical law over the full working extent of its parameter(s). Proper construction can mean the benefit of scale ranges limited to values for which the function is valid (Giesecke et al. 1967).

It will doubtless be argued that developments such as touch-screen technology make eventual computer supremacy inevitable. The reality is that design involves much parallel processing – of the mental variety. This occurs to best effect when all essential inputs (including the computer screen) are to hand simultaneously. The lone PC has yet to emulate this situation.

A chart format offering unrivalled readability and resolution is the three-parallel-scale nomogram. It is generally thought of as being restricted to the display of functions having the form $f(w) = f(u) +/- f(v)$ – although this evidently extends to dealing with $f(w) = f(u)f(v)$ by the simple expedient of taking logarithms.

The availability of the computer for chart construction opens up the possibility of wider use ['though not as handled in the 'definitive' text on the subject (Levens 1965)]. Provided the solution to be displayed, w, is a proper function, $w(u,v)$, of two parameters, u, v, nomogram representation[1] is a possibility even when determination of individual values of w involves lengthy computation. The economies of preserving the entire map of computed w-values in chart form are then beyond question.

A dual opportunity arises:

to re-visit and up-date Levens' defective account
   to apply the revised method to displaying a function whose potential in a specialist area of design has so far been overlooked.

The under-utilized function is introduced first.

---

[1] The range of the function to be displayed is that over which it increases or decreases monotonically. The restriction is eased if it is acceptable to have separate displays covering different ranges.

Stirling Cycle Engines: Inner Workings and Design, First Edition. Allan J Organ.
© 2014 John Wiley & Sons, Ltd. Published 2014 by John Wiley & Sons, Ltd.

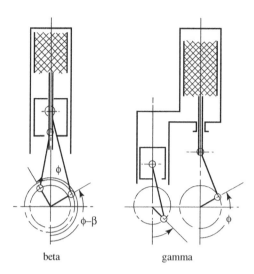

Figure A.3.1 The volume excursions and relative phase in a gamma engine can be identical to those of the beta type. On the other hand, strokes in the beta overlap, eliminating the 'additional dead-space' of the gamma

A.3.2 'Additional' dead space

Theoretical study of Stirling cycle machines starts with a definition of the cyclic variation of working-space volumes. There are obvious advantages to common algebra and coding, and to handling differences in basic layout by changes to input data.

Figure A.3.1 is a schematic representation of two of the principal layouts – the so-called 'beta' (coaxial) and 'gamma'. Even with piston motion rationalized to simple-harmonic, deriving the expression for compression-space variations in the 'beta' case is algebraically tricky because of the overlap of piston and displacer excursions – a feature which the 'gamma' arrangement avoids.

Finkelstein (1960) offers an expression for the additional dead (unswept) volume $V_{dc}{}^+$ incurred by gamma over beta. As a fraction of expansion space displacement V_E this is:

$$V_{dc}{}^+ / V_E = \tfrac{1}{2}\{1 + \lambda - \sqrt{(1 + \lambda^2 - 2\lambda \cos \beta)}\} \qquad (A.3.1)$$

In Equation A.3.1 parameter β is kinematic phase angle and λ is kinematic displacement ratio. The latter is the ratio of the excursion envelope of the compression piston to that of the displacer.

For the original purpose (quantifying the dead-space handicap of the 'lambda' engine) Equation A.3.1 is of limited interest, the more so following recent realization that increasing dead-space is not always a performance disadvantage – and sometimes the opposite. Its value is now seen to lie in allowing the simple expression for the volume variations of the gamma type to serve both machines *and handling the beta option by merely subtracting the numerical value of $V_{dc}{}^+$ from the existing compression-end dead space of the equivalent gamma.*

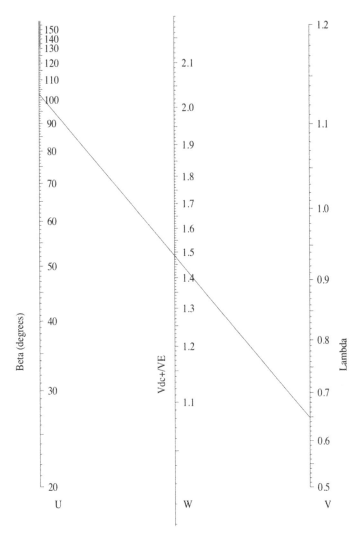

Figure A.3.2 Nomogram representation of Equation A.3.1. The diagonal line is a sample solution for specimen parameter values $\lambda = 0.65$, $\beta = 102$ degrees

Equation A.3.1 does not lend itself to rectification by taking logarithms. On the other hand, pre-processing by *anamorphosis*[2] results in the nomogram included here as Figure A.3.2

For the parameter values of the caption of Figure A.3.2 the hand calculator returns 1.4755 for V_{dc}^{+}/V_{E}. From the nomogram the reading is 1.48 or so. The close agreement conceals an aspect of chart construction requiring exploration if potential [and limitation(s)] are to be appreciated.

---

[2]A distorted projection or drawing … which appears normal when viewed from a particular point or by means of a suitable mirror.

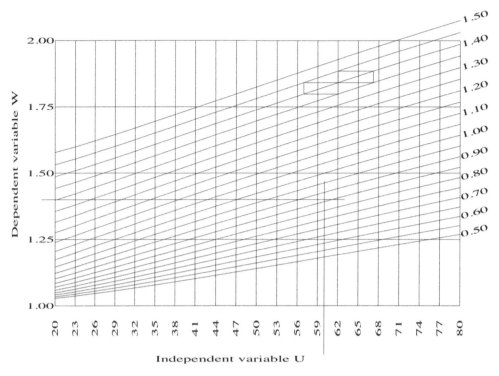

Figure A.3.3 Conventional concurrency (cartesian) plot of Equation A.3.1. Bilineality construction at upper right. Because the curves are almost parallel in this region the test gives the impression of closing. Printing out coordinates confirms that it does not

A.3.3 Anamorphosis and rectification

Figure A.3.3 is Equation A.3.1 in familiar cartesian representation. This and the nomogram of Figure A.3.2 are evidently mutual transformations, since they depict the same information.

A step in transforming from *x*–*y* cartesian (concurrency chart) to nomogram is *rectification*. Levens illustrates the rectification of two adjacent curves. He admits that … *it does not necessarily follow, however, that the other curves of the family will be rectified by the new scales x′ and y′. Wertheimer described a test for 'bilineality' which, if satisfied by three adjacent curves having a common point, would assure rectification of the curves by two new scales such as x′ and y′.*

A trial application of the test is illustrated on Figure A.3.3. For each set of three curves to have the necessary 'closure property', the rectangular chain should form a closed loop centred on a point of the inner curve. The curve family *happens to be* almost parallel, giving the impression that closure is achieved. Printing out the coordinates (an option not available to Levins) confirms otherwise. In one of the two constructions illustrated by Levens to make his point the test fails spectacularly! Remarkable in the circumstances is the implied expectation that the test *might* prove satisfactory: behind most families of curves are uniform (or arbitrary) increments in the *numerical value of the parameter*. Except in the special case of curves which

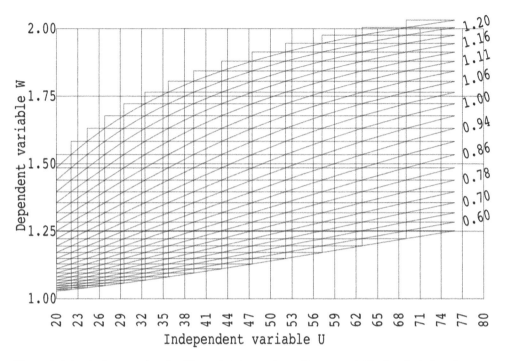

Figure A.3.4   Earlier cartesian plot (Figure A.3.3) transformed by sequential back-calculation of numerical values for parameter *v* (sloping characters at right) so that they pre-determine satisfactory outcome of the bilineality condition at all locations

are already straight lines, closure of the rectangular loop calls for a particular *physical spacing* between adjacent curves. Only by chance is spacing in proportion to parameter value.

Closure can, however, be achieved by prior choice of appropriate curve spacing. Corresponding parameter values can then be back-calculated. Figure A.3.4 is the result for Equation A.3.1. Note that parameter values (upper right-hand side) are not those of Figure A.3.3.

The entire map is now a superposition of coincident bilineality constructions, all of which evidently close. Anamorphic transformation is simultaneously achieved to parameter scales u' and v' whereby a specimen numerical solution value (i.e., of $V_{dc}{}^{+}/V_E$ in this case) corresponds to a *range* of known values of the new parameters u' and v'.

It is the cartesian concurrency chart *in this form* – and not the original – which lends itself by a further anamorphic transformation to standard form $f(w) = f(u) +/- f(v)$ and thence to nomogram format. This final step is elementary, merely requiring that dependent variable w be mapped into w', where successive increments in w' are uniform.

The reworking of Levens' treatment is not yet complete: on Figure A.3.4 u' and v' are *locations* of values of u and v rather than values of u and v themselves. At every location (intersection of u' and v' points) a numerical value can be calculated for w from $w = w(u, v)$ – in the present case from Equation A.3.1. When function w is of the form $f(w) = f(u)f(v)$, then the w values are precisely those of the construction used to locate the intersection points. Equation A.3.1 is not of this simple form, and discrepancies arise. These do not represent round-off or truncation: they are intrinsic to the construction. Magnitude varies according to the function being mapped, is a minimum at the left of the figure (where rectification starts) and compounds towards the right. The peak discrepancy arising on Figure A.3.4 is some $3\frac{1}{2}\%$. It is the transformed values which embody the error, which thus carries over to the nomogram. What saves the day (and the method) is that the error can be calculated point by point as construction proceeds, and the project discontinued if the result would be unserviceable.

A.3.4 Post-script

If this account is of value, the major benefit is about to be revealed: the writer first came into contact with graphical methods in general, and with the nomogram in particular, in about 1965. The design of nomograms remained a sideline interest until, in the unlikely setting of a market in the Brazilian city of Belo Horizonte, a copy of Levens' 1965 book, with price pencilled in pounds (£2.00), was acquired for a fraction of that price in Cruzeiros. The acquisition re-ignited interest: here apparently, was a way of generalizing the handling of functions of two variables $w = w(u,v)$ – analytical *or experimental* (pp. 119–121). Moreover – to quote Levens at p 121 – '*The nomographic method provides an excellent tool for checking the validity of the family of experimental data curves*'.

Attempts at realizing these benefits have occupied this writer over the subsequent 37 years – on and off (obviously) but right up to the date of this writing. The 'breakthrough' came as recently as the final week of March, 2012, when the only option remaining was to discard Levins approach and to begin again from scratch.

The epiphany of itself is minor. The very occurrence does, however, chime with a far grander context – Koestler's scholarly and life-changing account of the evolution of modern scientific understanding and method. One of several unifying threads is summarized in the Wikipedia entry for Koestler:

> Another recurrent theme of this book is the breaking of paradigms in order to create new ones. People – scientists included – hold onto cherished old beliefs with such love and attachment that they refuse to see the wrong in their ideas and the truth in the ideas that are to replace them.

The major benefit of the present account? The link to the Koestler book which no responsible scientist or engineer can afford not to read.

Should interest in nomography take hold, readers can steal a march on yours truly by accessing the masterly exposition by Doerfler (2009), which came to light after this present account was written, by accessing *http://www.myreckonings.com*. The site reviews a title by Evesham (2010), which sounds sufficiently indispensable to be cited here (see Bibliography) before this author has had the chance to acquire a copy. Doerfler is evidently no ordinary mathematician, and generously makes his e-mail address available (see his site) for exchanges on his wide-ranging interests.

At a lighter level, readers new to the subject might find the account by Earle (1977) more accessible than the specialist sources.

Appendix 4

Kinematics of lever-crank drive

At one stage the author was under the illusion of having invented this mechanism. An application was submitted for a United Kingdom patent, since when the earlier patent by James Wood of Sunpower has come to light.

In Figure A.4.1 the crankshaft rotates clockwise about point O, driving crank-pin P, offset from the crankshaft axis by radial distance r. Crank angle φ is measured clock-wise from the upper vertical axis. Lever S-R-Q is pivoted at ground point Q. Bell-crank P-R-T pivots at crank-pin P and point R on the lever. Points O-P-R-Q define the classic 'four-bar linkage'. Point S may drive the piston and T may drive the displacer – or T the piston and S the displacer, reversing the direction in which the cycle work is positive.

Applying Pythagoras' theorem:

$$c/r = \sqrt{[(X/r)^2 + (Y/r)^2]}$$

The fixed angle ξ is expressed in terms of the horizontal and vertical coordinates of Q relative to O:

$$\xi = \operatorname{atan}(X/Y)$$

Point S is uppermost at the value of φ for which O-P-R (in that order) form a straight line, and lowermost when P-O-R (in that order) are aligned. These conditions determine the dead-centre (inner/outer) positions of the member (piston or displacer) connected at S.

By the cosine formula applied to triangle O-R-Q with O-P-R in alignment:

$$\psi_{\text{IDCs}} = \operatorname{acos}\{\tfrac{1}{2}[RQ^2 + c^2 - (r+H)^2]/(RQ+c)\}$$
$$= \operatorname{acos}\{\tfrac{1}{2}[(RQ/r)^2 + (c/r)^2 - (1+H/r)^2]/(RQ/r + c/r)\}$$

Similarly:

$$\psi_{\text{ODCs}} = \operatorname{acos}\{\tfrac{1}{2}[(RQ/r)^2 + (c/r)^2 - (1-H/r)^2]/(RQ/r + c/r)\}$$

Stirling Cycle Engines: Inner Workings and Design, First Edition. Allan J Organ.
© 2014 John Wiley & Sons, Ltd. Published 2014 by John Wiley & Sons, Ltd.

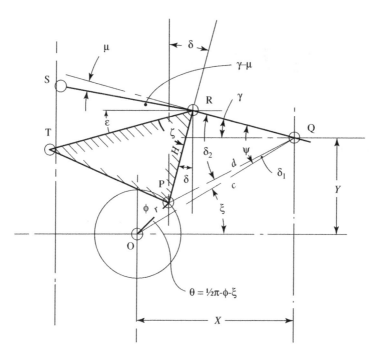

Figure A.4.1   Notation (_to no scale!_) for analysis of a quasi-linear 'lever-crank' drive mechanism offering thermodynamic flexibility combined with minimal side thrust. For viable operation $H + RQ > r + c$

The dead-centre values of angle γ follow by simple subtraction:

$$\gamma_{IDCs} = \psi_{IDCs} - \xi$$

$$\gamma_{ODCs} = \psi_{ODCs} - \xi$$

With crank angle φ measured from the uppermost position in the figure:

$$\theta_{anti\text{-}clock} = \tfrac{1}{2}\pi + \varphi - \xi$$

$$\theta_{clockwise} = \tfrac{1}{2}\pi - \varphi - \xi$$

By a further application of the cosine formula:

$$d^2 = c^2 + r^2 - 2cr\cos\theta$$

$$d/r = \sqrt{[(c/r)^2 + 1 - 2(c/r)\cos\theta]}$$

An expression for auxiliary angle δ_1 may be found by considering the right-angled triangle formed by sides P-Q, $r\sin\theta$, and $c - r\cos\theta$:

$$\delta_1 = \operatorname{atan}[r\sin\theta/(c - r\cos\theta)]$$

$$= \operatorname{atan}[\sin\theta/(c/r - \cos\theta)]$$

The variable angle ψ is the sum of angles δ_1 and δ_2. This last follows from a further application of the cosine formula to the triangle formed by sides H, $R\text{-}Q$ and d:

$$H^2 = RQ^2 + d^2 - 2dRQ\cos\delta_2$$
$$\delta_2 = \mathrm{acos}\{\tfrac{1}{2}[(RQ/r)^2 + (d/r)^2 - (H/r)^2]/(d/r \cdot RQ/r)\}$$
$$\psi = \delta_1 + \delta_2$$

Vertical and horizontal excursions of point S are available in terms of angle γ:

$$\gamma = \psi - \xi$$
$$x_R/r = e/r + X/r - (QR/r)\cos\gamma$$
$$y_R/r = Y/r - (QR/r)\sin\gamma$$
$$x_S/r = x_R/r - (RS/r)\cos(\gamma - \mu)$$
$$y_S/r = y_R/r - (RS/r)\sin(\gamma - \mu)$$

The considerable kinematic versatility of the mechanism is appreciated on noting:

that rotation may be clockwise or anti-clockwise – with a different sequence of volume variations in the two cases;

that the piston may be connected at T, the displacer at S – or the opposite way around, again with different respective variations of volume with crank angle φ;

that the cylinder may be situated <u>below</u> the mechanism ('inverted' relative to the usual orientation) or <u>above</u>.

Let

$$z = X - RQ\cos\gamma + r\sin\varphi \quad \text{anti-clockwise rotation;}$$
$$= r\sin\varphi - (X - RQ\cos\gamma) \quad \text{clockwise rotation.}$$
$$\delta = \mathrm{asin}(X \cdot z/H)$$
$$\zeta = \varepsilon + \delta + \tfrac{1}{2}\pi = \pi$$

Angular inclination ε of side $R\text{-}T$ to the horizontal is:

$$\varepsilon = \zeta + \delta - \tfrac{1}{2}\pi \quad \text{anti-clockwise rotation}$$
$$= \tfrac{1}{2}\pi - \zeta - \delta_1 \quad \text{clockwise rotation}$$

Recalling that horizontal distances originate at point O:

$$x_T/r = x_R/r - (RT/r)\cos\varepsilon$$
$$y_T/r = y_R/r - (RT/r)\sin\varepsilon$$

References

Anon (no date) NHS Estates Health Technical Memorandum HTM 2025 Ventilation in health care premises. Pt. 2 Design considerations ISBN 0113217528.

Anon (no date) SIMQX – appears in documentation for IBM 1130 Scientific Subroutine Library.

Anon 1966 Mechanical and physical properties of the austenitic chromium-nickel stainless steels at elevated temperatures. International Nickel Ltd., Thames House, Millbank, London (address at 1966).

Anon 1968 Properties of some metals and alloys, Ed. 3. International Nickel Co. Inc, 67 Wall St., New York, NY 10005 (address at 1968).

Anon (*circa* 1977) Stirling cycle engine 1817–1977. Owner's manual for engine model # 1, Solar Engines Inc. 2937 W Indian School Rd., Phoenix, AZ 85017.

Anon 1981 The properties of aluminium and its alloys, Ed. 8 (amended) Aluminium Federation (ALFED) Broadway House, Birmingham B15 1TM (address at 1981).

Anon 2005 Referee report on behalf of the editor of *Cryogenics* on proposed paper 'The friction factor – an impediment to regenerator design'. Also subsequent report by same referee on re-submission modified to requirements of original report.

Anon 2006 Web site of Recemat International 'Recemat® metal foam'. URL: http://www.recemat.com.

Adler A A 1946 (Feb) Variation with Mach number of static and total pressures through various screens. National Advisory Committee for Aeronautics (NACA) Wartime Report # L5F28.

Annand W J D 1963 Heat transfer in the cylinders of reciprocating internal combustion engines. Proceedings, Institution of Mechanical Engineers **177** (36) pp 973–990.

Archibald J 2006 Communication with the author by e-mail.

Barree R D and Conway M W 2004 Beyond beta factors: a complete model for Darcy, Forchheimer and trans-Forchheimer flow in porous media. Paper 89325. Proceedings, Annual Technical Conference, Society of Petroleum Engineers.

Bird R B, Stewart W E and Lightfoot E N 1960 *Transport phenomena* John Wiley and Sons, New York, NY.

Bomford N 2004 Communication with the author (3 January).

Boomsma K et al. 2003 Simulations of flow through open cell metal foams using an idealized periodic cell structure. International Journal of Heat and Fluid Flow. **24** pp 825–834.

Bopp G and Co (undated) Woven wire cloth. Document G 0100/02, G Bopp AG, Zurich.

Bowden C E 1947 *Diesel model engines*. Percival Marshall, London.

Bryson W 2003 *A short history of nearly everything*. Doubleday, London.

Bumby J R and Martin R 2005 Axial flux permanent magnet generators for engine integration. Proceedings 12th International Stirling Engine Conference, University of Durham, pp 463–471.

Cambell Tools Co 2005 Website of Campbell Tools Company Springfield, OH 45504.

Carnot S 1824 *Réflexions sur la puissance motrice due feu et sur les machines propres a développer cette puisance*. Bachelier, *Paris*. Re-printed in Annales Scientifiques de l'É.N.S. 2e série, tome 1 (1872) pp 393–347.

Chapman A J 1967 *Heat transfer*. Collier-MacMillan, London.

Chen Z et al. 2001 (Aug) Derivation of the Forchheimer law *via* homo-genization. Transport in Porous Media **44** # 2 pp 325–335.

Stirling Cycle Engines: Inner Workings and Design, First Edition. Allan J Organ.
© 2014 John Wiley & Sons, Ltd. Published 2014 by John Wiley & Sons, Ltd.

CNW (undated) Precision engineering of stainless steel small tubes. Product catalogue of Coopers Needle Works Ltd., 261/265 Aston Lane, Birmingham B20 3HS.

Coppage J E 1952 The heat transfer and flow friction characteristics of porous media. PhD dissertation, Department of Mechanical Engineering, Stanford University.

Coppage J E and London A L 1956 Heat transfer and flow friction characteristics of porous media. Chemical Engineering Progress, 52 #2 pp 57F–63F.

CSC (undated) Illustrated catalogue of ceramic products. Ceramic Substrates and Components Ltd., Lukely Works, Carisbrooke Rd., Newport, Isle of Wight.

Dando R 1997 Selected details of CRE Stirling engine supplied by communication with the author.

Darcy H 1856 Les Fontaines Publiques de la Ville de Dijon, Dalmont, Paris.

De Brey H, Rinia H and van Weenen F L 1947 Fundamentals for the development of the Philips air engine. Philips Technical Review 9 # 4 pp 97–124.

Denney D 2005 Beyond beta factors: a model for Darcy, Forchheimer and trans-Forchheimer flow in porous media. Highlights for *JPT-online* from paper of similar title by Barree R D and Conway M W for Society of Petroleum Engineers (SPE) annual technical conference and exhibition, Houston, 26–29 September 2004.

Devois J F and Durastanti J F 1996 Modelling of a spiral plate heat exchanger by a finite difference method. Proceedings of Conference: New developments in heat exchangers, Edited by N Afgan et al. ISBN 905699512X Part 5 (Plate-type exchangers) Gordon and Breach, Amsterdam pp 377–386.

Doerfler R 2009 On jargon – the lost art of nomography. Jnl. UMAP 30, 4, 457–493.

Dunlop Co plc (undated) Retimet® metal foam. Manufacturer's sales brochure, Dunlop Ltd. Aviation Division – Equipment, Holbrook Lane, Coventry CV8 4QY.

Dunn A R 1980 Selection of wire cloth for filtration/separation. Re-printed from *Filtration and Separation*, Sept./Oct. issue. (Pages of re-print un-numbered.)

Earle J H 1977 *Engineering design graphics* Addison-Wesley, Reading Massachusetts (3rd edn, second printing).

Ergun S 1952 Flow through packed columns. Chemical Engineering Progress 48, pp 89–94.

Evesham H A 2012 *The history and development of nomography*. Docent Press, London.

Faires V M 1969 *Design of machine elements* Collier-MacMillan, New York.

Feulner P 2013 e-mail communication recounting experiments with single-acting V-engine at Esslingen (1997/8) and exchanges with Henrik Carlsen (30 April).

Finkelstein T 1960a Optimization of phase angle and volume ratio for Stirling engines. Paper 118C Proceedings, SAE Annual Meeting, Detroit, Michigan, January 11–15.

Finkelstein T 1960b Generalized thermodynamic analysis of Stirling engines. Paper 118B Proceedings, SAE Annual Meeting, Detroit, Michigan, January 11–15.

Finkelstein T and Organ A J 2001 *Air Engines* Originally published by Professional Engineering Publishing, London and Bury St. Edmunds. (2005 onwards: John Wiley and Sons, Chichester).

García-Granados FJ 2013 Software for the analysis of the Stirling cycle: GGSISM. http://www.stirlinginternational. org/e-journal/item.asp?id=B2401EC0-E890-4B50-AB0E-A34AEFB08EFD.

Gedeon D 1989 Modelling 2-D jets impinging on Stirling regenerators. Proc. 24 IECEC 5 2199–2203.

Giesecke F E Mitchell A Spencer H C and Hill I L 1967 *Technical drawing*. Collier-Macmillan, London.

Gifford W E and Longsworth R C 1964 Pulse-tube refrigeration. Transactions of the American Society of Mechanical Engineers (ASME) August pp 264–268.

Granta Design 2006 Web site http://www.grantadesign.com.solutions.

Gunston W 1998 *The Life of Sir Roy Fedden*. Rolls Royce Heritage Trust, London.

Heaton H S Reynolds W C and Kays W M 1964 Heat transfer in annular passages. Simultaneous development of velocity and temperature fields in laminar flow. International Journal of Heat Mass Transfer. 7 pp 763–781.

Hargis A M and Beckmann A T 1967 Applications of spiral plate heat exchangers. Chemical Engineering Progress. 63, # 7 pp 62–67.

Harness J B and Neumann P E L 1972 A theoretical solution of the shuttle heat transfer problem. Proceedings of International Conference on Cryogenic Engineering 16 pp 97–100.

Harvey J P 2003 Oscillatory compressible flow and heat transfer in porous media – application to cryo-cooler regenerators. PhD dissertation, Georgia Institute of Technology, GA.

Hausen H 1929 Wärmeaustausch in Regeneratoren. Zeitschrift des Vereins deutscher Ingenieure 73, p 432.

Hislop D 2006 Communications with the author (e-mail and telephone) throughout the year.

Hodgman C D (ed.) 1954 *Mathematical tables from Handbook of Chemistry and Physics*. Chemical Rubber Publishing Co., Cleveland, Ohio (fifth printing 1957).

Igus UK 2006 Technical sales information on polymer bearing bushes and related products. (Web-site readily located by Googling 'Igus').

IMI/Amal (undated) Flame arresters. Illustrated trade catalogue published by Amal, division of IMI, Holdford Rd., Witton, Birmingham B6 7ES.

Inco Ltd 1966 Mechanical and physical properties of the austenitic chromium-nickel stainless steels at elevated temperatures. INCO Alloys International, Hereford UK.

Jakob M 1957 *Heat transfer II*. Wiley, New York. Also Chapman and Hall, London.

Johnson R C 1971 *Mechanical design synthesis with optimization applications*. Van Nostrand Reinhold, New York.

Karabulut H et al. 2009 An experimental study on the development of a beta-type Stirling engine for low and moderate heat sources. Applied Energy **86** pp 68–73.

Kays W M and London A L 1964 *Compact Heat Exchangers*. McGraw Hill, New York.

Koestler A 1959 *The Sleepwalkers: A History of Man's Changing Vision of the Universe* (1959), Hutchinson, London.

Kolin I 1991 *Stirling motor* Zagreb University Publications. (Contributions to 5th Int. Stirling Engine Conf., Inter-University Centre, Dubrovnik, 1991).

Liadlaw-Dickson D J 1946-7 *Model diesels*. Harborough Publishing, Leicester.

Larque I 1998 Communication with the author.

Larque I 2002 Design and test of a self-pressurized high compression-ratio Stirling air engine. Proceedings of European StirlingForum, Osnabrück, Published on CD-rom by ECOS Gesellschaft für Entwicklung und Consulting Osnabrück mbH, Westerbreite 7 D49084 Osnabrück, Germany.

Laws E M and Livesey J L 1978 Flow through screens. Annual review of fluid mechanics **10**, 247–266 (January).

Levens A S 1965 *Graphical methods in research*. Wiley, New York.

Lighthill J 1975 *Mathematical BiofluidDynamics*. Society for Industrial and Applied Mathematics, Philadelphia, PA.

Lo C-B 1999 Appraisal of a 25cc Stirling cycle engine. MPhil dissertation, University of Cambridge.

Ludwig H 1964 Experimentelle Nachprüfung der Stabilitätstheorien für reibungsfreie Stromungen mit schraubenlinienformigen Stromlinien. Zeitschrift Flugwissenschat **12** Heft 8.

Magara Y et al. 2000 Effects of stacking method on heat transfer characteristics in stacked wire gauze. Proceedings of 4th Symposium on Stirling cycle, Japan Society of Mechanical Engineers (JSME) pp 43–46.

Marshall A 1994 *The Marshall Story: a century of wheels and wings*. Sparkford; P. Stephens.

Martini W M 1978 Stirling engine design manual. US DOE/NASA report. Washington, DC.

Miyakoshi R et al. 2000 Study of heat transfer characteristics of regenerator matrix (2nd report). Proceedings, 4th symposium on Stirling cycle, Japan Society of Mechanical Engineers.

Model Engineering Co 1974 Correspondence ref. SWI/ABM from the company to the author.

Moin P and Kim J (date not available) Tackling turbulence with supercomputers. Scientific American at http://turb.seas.ucla.edu/~jkim/sciam/ turbulence.html.

Morgan P G 1959 (Aug) High speed flow through wire gauzes. Journal of the Royal Aeronautical Society **63**.

Naso V 2003 Chairman's introduction, call for papers, 11th International Stirling engine conference, Rome, 19–21 November.

Niu Y Simon T W Ibrahim M B Tew R and Gedeon D 2003 JET Penetration into a Stirling Engine Regenerator Matrix with Various Regenerator-to-Cooler Spacings, Paper # AIAA-2003-6014, Presented at the 2003 International Energy Conversion Engineering Conference, Portsmouth VA.

Ochoa F Eastwood C Ronney P D and Dunn B 2003 Thermal transpiration based on micro-scale propulsion and power generation devices. Proceedings, 7th International Microgravity Combustion Workshop, Cleveland, Ohio (June – 4pp un-numbered).

Organ A J 1992 *Thermodynamics and Gas Dynamics of the Stirling cycle machine*. Cambridge University Press, Cambridge.

Organ A J 1992 'Natural' coordinates for analysis of the practical Stirling cycle. Proceedings, Institution of Mechanical Engineers Pt. C **206** pp 407–415.

Organ A J 1995 The hot-air engine – a very un-scale model. Paper read at the annual Model Engineering Exhibition, Olympia, London.

Organ A J and Maeckel P 1996 Connectivity and regenerator thermal shorting. Proceedings, European Stirling Forum, pp 229–243, Fachhochschule Osnabrück.

Organ A J 1997 *The regenerator and the Stirling engine*. PEP (Professional Engineering Publishing), London and Bury St. Edmunds. (2005 onwards: John Wiley and Sons, Chichester).

Organ A J 2000 Regenerator analysis simplified. Proceedings European Stirling Forum Osnabrück (22–24 February) pp 27–37.

Organ A J 2001 *Stirling and pulse-tube cryo-coolers*. John Wiley and Sons, Chichester.

Organ A J and Lo C-b 2000 Thermodynamic personality of a 25cc Stirling engine. Pt. II: Hydrodynamic pumping loss and thermal short. Proceedings, European Stirling-Forum, Osnabrück pp 49–55 (February).

Organ A J and Larque I 2004 A recuperative combustion chamber for a small air engine. Proceedings, European Stirling-Forum, Osnabrück, Published on CD-rom by ECOS Gesellschaft für Entwicklung und Consulting Osnabrück mbH, Westerbreite 7, D49084 Osnabrück, Germany.

Organ A J 2004 Solution of the 'shuttle heat transfer' problem. Proceedings, International Stirling-Forum, Osnabrück. Published on CD-rom by ECOS Gesellschaft für Entwicklung und Consulting Osnabrück mbH, Westerbreite 7, D49084 Osnabrück, Germany.

Organ A J 2005a *Stirling and pulse-tube cryo-coolers*. John Wiley and Sons, Chichester.

Organ A J 2005b The flow friction correlation – an obstacle to regenerator design. Conditionally accepted for publication in *Cryogenics*, Elsevier, London.

Organ A J 2005c Regenerator thermal analysis – un-finished business. Under consideration for possible publication in Proceedings, Japan Society of Mechanical Engineers (JSME).

Organ A J 2005d Regenerator design and the neglected art of back-of-the-envelope calculation. Under review for publication in Proceedings, Institution of Mechanical Engineers, Pt. C.

Oswatitsch K 1956 *Gas dynamics*. English version by Gustav Kuerti, Academic Press, New York p 60.

OUP 1993 New Shorter Oxford Dictionary, Oxford University Press.

Pinker R A and Herbert M V 1967 Pressure loss associated with compressible flow through square-mesh wire gauzes. Proceedings, Institution of Mechanical Engineers, Pt. C (Journal of Mechanical Engineering Science) **9**, pp 11–23.

Rinia H and DuPré F K 1946 Air engines. Philips Technical Review **8** #5, pp 129–160 (May).

Rios P A 1971 An approximate solution to the shuttle heat transfer losses in a reciprocating machine. Journal of Engineering for Power, American Society of Mechanical Engineers (ASME), April, pp 177–182.

Rix D H 1984 An enquiry into gas process asymmetry in Stirling cycle machines. PhD dissertation, University of Cambridge.

Rollet H & Co. 2005 Manufacturer's web site. Postal address Dawley trading Estate Stallings Lane Kingswinford West Midlands DY6 7BW.

Ross M A 1989 States-side hot air. Model Engineer **162**, # 3842 pp 155–156.

Ruehlich I and Quack H (undated) Investigations on regenerative heat exchangers. Technical University, Dresden.

Schmidt G 1871 Theorie der Lehmann'schen calorischen Maschine. Zeitschrift des Vereins deutscher Ingenieure, Band XV, Heft 1 (January).

Schneider P J 1955 *Conduction heat transfer*. Addison-Wesley, Reading, Massachusetts.

Schumann T E W 1929 Heat transfer in a liquid flowing through a porous prism. J. Franklin Inst. **208**, 1243–1248 (104th year).

Sier R 1995 *The Revd. Robert Stirling DD*. L A Mair, Chelmsford.

Sier R 1999 *Hot Air, Caloric and Stirling Engines* – I: A History. L A Mair, Chelmsford UK.

Shapiro A H 1954 *The dynamics and thermodynamics of compressible fluid flow*. Ronald, New York, **I** p 184, **II** p 1131.

Simmons W F and Cross H C 1952 Report on the elevated-temperature properties of stainless steels. Special Technical Publication # 124, American Society for Testing Materials (ASTM).

Stirling J 1845 Description of Stirling's improved air angine. Proceedings, Institution of Engineers **IV** pp 348–361.

Stirling J 1853 Contribution (at pp 599–600) to discussion 'Heated Air Engine'. Proceedings, Institution of Civil Engineers **XII** 558–600.

Stirling R 1816a Orthographic drawings attributed to London patent 4081: '*Improvements for diminishing the consumption of fuel, and in particular an engine capable of being applied to the moving of machinery on a principle entirely new*'.

Stirling R 1816b Edinburgh patent 4081 (as above, but with superior illustrations).

SKF Ltd 2002 SKF bearings with *solid oil* – the third lubrication choice. Publication 4827/1 E, SKF Corporation, Sweden.

Su C-C 1986 An enquiry into the mechanism of pressure drop in the regenerator of the Stirling cycle machine. PhD dissertation, Engineering Department, Cambridge University.

Tailer P L 1995 Thermal lag machine. US Patent # US5414997.

Taylor G I 1935 Distribution of velocity and temperature between concentric rotating cylinders. Proceedings of the Royal Society of London, series A **135** pp 494–512.

Taylor G I 1936 Fluid friction between rotating cylinders. I – Torque measurement. Proceedings of the Royal Society of London series A, **157** pp 546–564.

Tenmat 1996 Illustrated catalogue of high temperature engineering boards. Tenmat Snc. Z.A.I. (RN19) 77370 Nangis, France. UK contact: Tenmat Midland Ltd.

Thompson B E (undated) Evaluation of advanced heat recovery systems. Web page of Foam Application Technologies Inc.

Thwaites B 1949 Approximate calculation of laminar boundary layer. Aeronautical Quarterly **1**, pp 245–280.

Urieli I and Berchowitz D 1984 *Stirling cycle engine analysis*. Adam Hilger, Bristol.

van Rijn C 2005 Report on work carried out at Cambridge University Engineering Department as part of engineering degree programme at the University of Eindhoven, Netherlands.

van Rijn C 2006 Interim account of further work at Cambridge University Engineering Department towards engineering degree programme at the University of Eindhoven, Netherlands.

Walker G 1980 *Stirling engines* Oxford University Press.

Warbrooke E T 2005 *Building Stirling-1*. Camden Miniature Steam Services, Rode, Somerset.

Ward G 1972 Performance characteristics of the Stirling engine. MSc dissertation, University of Bath.

Weisstein E W 2005 Cubic Close Packing, from MathWorld – A Wolfram Web Resource. http://mathworld.wolfram.com/CubicClosePacking.html.

Wellington Engineering 2006 Web site of Parkside House, Rigby Lane, Hayes, Middlesex, UB3 1ET, UK, Tel: (0)20 8581 0061, www.welleng.co.uk.

West C D 1993 Some single-piston closed-cycle machines and Peter Tailer's thermal lag engine. Paper 93105, Proceedings, 28th Inter-Societies Energy Conversion Engineering Conference (IECEC), American Chemical Society, pp 2.673–2.679.

White F M 1974 *Viscous fluid flow*. McGraw-Hill, New York, p 206.

Wicks F and Caminero C 1994 Peter Tailer external combustion thermal lag piston/cylinder engine analysis and potential applications. Paper AIAA-94-3987-CP, Proceedings, Inter-Societies Energy Conversion Engineering Conference (IECEC) pp 951–954.

Wilson R S 2000 (25–26 Oct.) Advances in piston and packing ring materials for oil-free compressors. Presented to 4th Workshop on Piston Compressors, Kötter Consulting Enginers, Rheine, Germany.

Wirtz R A et al. 2003 Thermal/fluid characteristics of 3-D woven mesh structures as heat exchange surfaces. Transactions, Institution of Electrical and Electronic Engineers (IEEE) – components and packaging technologies **26** # 1, pp 40–47.

Name Index

Archibald J, 55, 56, 220

Bazan K, 7
Beale W T, 120
Bejan A, 22, 151
Berchowitz D, 55, 66, 199
Bomford N, 65
Bopp G and Co, 137
Britter R E, 47

Carnot S, 4, 9
Chapman A J, 187
Clapham E, 84, 85
Corey J, 46

Doerfler R, 256

Earle J H, 256
Evesham H A, 256

Feulner P, xviii, 37, 173, 176
Feynman R, 45, 46
Finkelstein T, xviii, 2, 25, 27, 34, 36, 37, 65,
    110, 120, 146, 168, 199, 224, 227, 252
Flynn G, 5

Gay Lussac, 64
Gedeon D, 177
Giesecke F E, 250

Hargreaves C M, 21, 105, 163–165
Hartford (Company), 5

Hausen H, 171
Herbert M V, 192
Hino M, 46
Hitchcock A, 20

Jakob M, 202
James R G, xviii
Jenkin H C Fleeming, 1
Joule J P, 6

Karabulut H, 33–34, 36, 37, 41, 42, 70
Kays W M, 102, 105, 177, 182, 195
Kim J, 45
Kirkley D W, 36
Koestler A, 265
Kolin I, 6, 171

Lamb H, 45, 46
Larque I, 6, 27, 32, 35–37, 41, 65, 70
Levens A S, 250, 255–256
Lighthill J, 200
London A L, 102, 105, 177, 182, 195

Maeckel P, xviii
Mead T, 5, 8
Miyabi H, 187, 188
Moin P, 45

Newton, I, 56
Niu Y, 177

Organ A J, 26, 36, 66, 199

Stirling Cycle Engines: Inner Workings and Design, First Edition. Allan J Organ.
© 2014 John Wiley & Sons, Ltd. Published 2014 by John Wiley & Sons, Ltd.

Subject Index

Stirling Cycle Engines: Inner Workings and Design, First Edition. Allan J Organ.
© 2014 John Wiley & Sons, Ltd. Published 2014 by John Wiley & Sons, Ltd.

Printed and bound by CPI Group (UK) Ltd, Croydon, CR0 4YY

16/04/2025

14658383-0001